ASHMIT RAJ

Physics of the Intrigued

Where Questions Become Universes

Scientific theories, facts, and models discussed throughout the book—such as those relating to quantum mechanics, relativity, cosmology, particle physics, and thermodynamics—are credited to their respective scientific communities and discoverers. This book does not claim to own or redefine any such established scientific knowledge.

This work is a philosophical and conceptual presentation aimed at making complex physics more accessible and engaging for readers. The interpretations and hypothetical frameworks are original creative expressions inspired by known scientific principles.

All efforts have been made to present ideas respectfully and in alignment with fair use principles for educational and transformative purposes.

If any individual or institution feels any part of the content requires additional credit or correction, they are encouraged to reach out directly to the author.

First edition

This book was professionally typeset on Reedsy.
Find out more at reedsy.com

To my parents, for their endless support.
To all the curious minds who look at the stars and wonder.
For those who dare to ask why.

Contents

Preface

This is not a textbook. -

Instead, what you hold is a journey through the most powerful force of all—curiosity.

Physics, to me, is not a subject. It is a lens. A way of looking at everything from galaxies tearing across the night sky to the silent mysteries within a vacuum. It is both the language and the logic of reality. And yet, even with all our equations and theories, there are still more questions than answers.

Physics of the Intrigued is my attempt to reframe the questions. Each chapter is a dive into a concept that has intrigued humanity—from the nature of time and observation to the edge of the observable universe and beyond. The content moves between classical clarity and quantum uncertainty, between scientific grounding and conceptual speculation.

I wrote this book for two kinds of readers:

The beginner, who feels awe but not always understanding.

The thinker, who has learned much but dares to wonder still.

This is not fiction, but it is philosophical. The theories inside are known, but the reflections are mine. You will find no math-heavy pages here in the main chapters, but you will find thought experiments, mental gymnastics, and a fair share of "what ifs."

I should also mention something important. I wrote this book as a 9th-grade student, currently in 10th grade. I am not a professional astrophysicist. Neither do I claim mastery over the complex theories discussed here nor the very things I'm currently studying. My goal was never to present myself as an expert, but rather as someone who sees physics not merely as a subject, but as a path of creativity, curiosity, and discovery.

So, whether you're -

A Student

A Dreamer

Or A Skeptic

—I welcome you.

Let's rethink what we know and unlearn what we assume.

"The universe is watching. Let's watch back."

— Ashmit

March 2026

Acknowledgments

I would like to express my heartfelt gratitude to my parents, Mr Chandan Mishra and Mrs Rupa Jha, for their constant encouragement, unwavering support, and belief in my abilities throughout the writing of this book. Their faith in my curiosity and passion for understanding the universe gave me the confidence to pursue this project and bring it to completion. They have always inspired me to ask questions, think deeply, and never lose my sense of wonder about the world around me.

I am deeply grateful to Rev. Thomas K. Thomas, the Principal of St. Thomas School Ranchi, for his continuous encouragement and strong support of my intellectual pursuits. His willingness to take an interest in my work, along with his thoughtful guidance and motivation, played a significant role in shaping my journey while writing this book. His encouragement not only strengthened my confidence but also inspired me to approach learning with greater curiosity and seriousness.

I would also like to extend my sincere thanks to Mrs Monika Sood for helping me connect with various physics professors, providing valuable exposure to scientific thinking.

I am profoundly inspired by the ideas and writings of great scientists and communicators such as Stephen Hawking, Albert Einstein, Richard Feynman, Brian Greene, and Carl Sagan, whose works have deeply influenced my perspective and continue to inspire my curiosity about the universe.

Various digital tools were also used for grammar checking, content verification, image generation, and evaluation during the writing process.

Finally, I would like to thank every curious reader who chooses to explore this book. It is my sincere hope that *Physics of the Intrigued* encourages deeper curiosity, inspires thoughtful questioning, and helps readers appreciate the beauty and mystery of the universe.

Introduction

Curiosity is the engine of physics. This book is not a textbook; it is an invitation — a short tour through the ideas that made me fall in love with the universe and the questions that keep me awake at night. Here you will find a blend of clear explanation, mental experiments, and a few speculative detours that try to make the strange feel discoverable rather than forbidding.

We begin with a simple provocation: what if our blindness to certain directions of reality is not a failure of intellect but a limit of geometry? In Chapter 1, "The Watcher," we use dimensional thinking to ask whether beings with access to more degrees of freedom would seem miraculous to us — and what that then tells us about structure, observation, and the limits of scientific imagination. From there, the book moves into time, entropy, and why the universe remembers the past but not the future; into relativity and the way gravity and motion shape the rate of time; and then across scales, from the tiniest particles to the grandest cosmic webs.

My aim is practical and personal. I want to make difficult ideas accessible without dumbing them down. You will see very little heavy math here; instead, you will meet thought experiments, visual analogies, and conceptual maps intended to build intuition. Expect questions more than answers, clarity more than formal proof, and the occasional "what if" that stretches the standard picture — including one original conjecture I outline toward the end.

This book grew from lectures, conversations, and the works of thinkers who taught me to ask better questions: Albert Einstein, Stephen Hawking, Richard Feynman, Brian Greene, and Carl Sagan — but it is written from the vantage of a curious ninth-grader, not a seasoned researcher. That perspective is intentional: sometimes a fresh, impatient curiosity sees useful corners the expert has already learned to ignore.

How to read this book: treat it as a map rather than a manual. You can follow the chapters in order — they are arranged to build from foundations (what is a dimension? what is entropy?) toward the more speculative — or you can jump to the topics that pique you most. Each chapter ends with a few reflective questions and a pointer to the next idea; use those to guide your own exploration.

If one promise underpins everything here, it is this: physics is a language for asking better questions. I hope that by the final page, you will not only understand more about time, space, and matter, but that you will also feel a little braver about asking the kinds of questions that move science forward.

— Ashmit

1

The Watcher

"Invisible to us, it watches from an abode beyond the Origin and the Void."

For years, humans have been regarded as the most dominant species among all present within this blue sphere suspended in the silent architecture of space. We predict the return of comets before they carve their arcs across the sky. We split the atom and uncover energies once thought unreachable. We decode the faint radiation left behind by the birth of the universe and call it the cosmic microwave background. We compress thought into circuits and memory into silicon.

In the shadow of such achievements, it becomes easy, perhaps too easy, to assume that awareness culminates here.

But dominance is meaningful only within the limits of comparison.

A creature confined to a flat surface would consider itself complete within its plane. It would measure length and width. It would master every available direction. Yet it would remain entirely unaware of height. If a hand passed above it, the creature would not perceive the hand as a whole. At most, it would witness inexplicable cross-sections appearing and vanishing without cause.

Its ignorance would not arise from stupidity. It would arise from dimensional confinement.

And so beneath the confidence of our species, there emerges a quiet intellectual unease, not rooted in fear, but in structure. If awareness is shaped by dimensional access, then our understanding of reality may be limited not by intelligence but by geometry.

Several questions arise over this point of peculiarity.

Who is it?

Where is it?

Why can we not see it?

These are not theological provocations. They are structural inquiries.

If there exists a being superior to us, not in moral virtue or emotional capacity, but in dimensional standing, then its invisibility would not be mystical. It would be geometric. Just as the flat creature cannot perceive height, we may be incapable of perceiving directions that lie beyond the three we inhabit.

Before we begin the search for such a presence, we must accept a necessary restraint. One does not investigate the existence of higher-dimensional objects without first understanding what a dimension is. To ask where such a being resides without defining residence itself is to pursue answers without constructing the framework required to hold them.

Therefore, the hunt must wait.

First, we must understand the scaffold of reality.

What is a dimension?

How does reality acquire structure?

What does it mean for existence to expand from point to line, from line to plane, from plane to volume?

And if progression continues, where does it lead?

Only after we have walked patiently through these questions can we return to the possibility of the Watcher.

Understanding From Scratch

For this task to be done, let me not be your spiritual guide but a guide to answer this with the efforts of great scientists in the devising of Multi-Dimensional Theory in Physics in a manner pellucid to you instead. If we attempt to define the term dimension in a simple and disciplined way, we may describe it as an independent parameter required to specify the position of any object, point, or event in space and time.

The word independent is crucial. A dimension represents a direction or variable that does not depend on another for its description. Each new independent parameter increases the descriptive freedom of a system.

Dimensions can be grouped into several broad categories:

- Spatial dimensions
- Temporal dimension
- Compactified or extra spatial dimensions
- Abstract dimensions

In principle, there is no strict limit to how many types of dimensions one may define. Any independent variable necessary to describe a system can be treated as a dimension. Energy levels in physics, states in computing, configuration spaces in mechanics, and probability spaces in mathematics can all be interpreted dimensionally.

However, for the purpose of answering the central questions raised earlier, we will restrict ourselves to four primary types. The reason for this limitation is methodological. We intend to move from zero dimensions to increasingly higher-dimensional frameworks to examine whether the existence of a structurally superior being is logically possible. Other dimensional classifications, although important in their own domains, are not essential for this particular investigation.

At this stage, we understand what a dimension is and how it may be classified. Before combining advanced theories such as M-Theory with our objective, we must carefully examine the four types mentioned above.

1. Spatial Dimensions

We inhabit what is commonly described as a three-dimensional spatial world.

These three dimensions are:

- Length
- Breadth
- Height

They are independent directions that allow us to specify the position of any macroscopic object in gravitational space, meaning at scales where classical physics applies, and quantum effects can be ignored.

Using three coordinates, typically labelled x, y, and z, we can precisely determine the location of an object. These are called spatial dimensions because they define extension in space.

2. The Temporal Dimension

Position alone does not fully describe reality. An event is not defined only by where it occurs, but also by when it occurs.

Time acts as an additional independent parameter that orders events. It allows processes to unfold and enables cause to precede effect. When time is combined with the three spatial dimensions, we obtain a four-dimensional framework.

It is important to state correctly that time is called the temporal dimension, not a spatial one. Together, the three spatial dimensions and one temporal dimension form what we call spacetime.

This unification was formalised by Sir Albert Einstein in his Special Theory of Relativity, where space and time were no longer treated as separate entities but as components of a single geometric structure. Events are described using four coordinates: three for space and one for time.

At this point, our observable universe is described as four-dimensional.

But modern theoretical physics suggests that this may not be the complete picture.

3. Compactified or Extra Spatial Dimensions

To understand compactified dimensions, we must briefly refer to String Theory.

According to String Theory, when we probe matter at scales far smaller than atoms, beyond electrons and quarks, the fundamental constituents of reality are not point-like particles but extremely small one-dimensional objects called strings. These strings vibrate continuously in different patterns.

Each vibrational mode corresponds to a different particle.

However, for the mathematics of String Theory to remain internally consistent, additional spatial dimensions are required. If calculations are performed in only four dimensions, the theory produces inconsistencies, including negative probabilities and mathematical instabilities.

To resolve this, the theory requires extra spatial dimensions beyond the familiar three.

These extra dimensions are believed to be compactified, meaning they are curled up at extremely small scales, possibly near the Planck length (approximately 1.6×10^{-35} meters). They are not extended like the dimensions we move through daily. Instead, they are tightly folded into complex geometrical shapes.

One such class of shapes is known as Calabi–Yau manifolds. The geometry of these compactified dimensions determines how strings vibrate. Different geometric configurations allow different vibrational patterns, and therefore different particle properties.

Figure 1.1: A Calabi-Yau Manifold

In most versions of superstring theory, a total of ten dimensions is required: nine spatial and one temporal. M-Theory, which attempts to unify various string theories, proposes eleven dimensions.

Because there are many possible Calabi–Yau geometries, each leading to different physical constants and particle behaviours, this framework suggests the possibility that multiple consistent versions of physics could exist. This idea contributes to the concept of the multiverse, where different universes may operate under different physical laws.

Whether or not the multiverse exists remains an open question. What matters for us is that higher-dimensional structure is not speculative fantasy. It arises naturally from mathematical necessity within advanced physical theories.

4. Abstract Dimensions

Finally, we consider abstract dimensions.

These are not physical directions in space or time. Instead, they are independent variables used to describe states, properties, or degrees of freedom within systems.

For example, the state of a quantum system can be represented in a high-dimensional mathematical space known as Hilbert space. In such cases, dimensions do not represent directions we can move through physically. They represent independent components required to describe the system completely.

Abstract dimensions do not exist as spatial or temporal extensions in physical reality. However, they are essential for modelling and analysing complex systems.

At this point, you are aware of the fundamental classification of dimensions, their structural roles, and the essential idea behind String Theory and M-Theory.

Now the progression can begin.

If zero dimensions represent pure positional confinement, and higher dimensions introduce increasing degrees of freedom, then the question we raised earlier returns with greater precision.

If dimensional hierarchy is real, and if reality structurally expands with each added degree of freedom, then where do we stand within that hierarchy?

And more importantly,

Is it logically impossible that something stands above us within it?

Counting On Fingers First

Consider a point.

No length.

No breadth.

No height.

It is not an object in the physical sense. It is a positional marker. A mathematical abstraction used to represent location without extension.

For this reason, it is said to be zero-dimensional.

Because a point has no size, a line of any length can contain infinitely many points. The line is continuous. Between any two points, another point can always be found. This is a property of mathematical continuity.

However, in physical reality, a perfect zero-dimensional object does not exist independently. Every measurable entity occupies some extent in space and persists through time. Pure geometric ideals exist precisely because mathematics allows exact abstraction, whereas nature operates within measurable limits.

At this stage, one might argue that time exists independently. One might claim that time existed before the Big Bang and will continue after the universe ends.

But this statement requires caution.

In modern cosmology, according to General Relativity, time as we understand it is intertwined with space. The Big Bang is not merely an explosion within space; it represents the origin of spacetime itself within our current physical model. Asking what happened before the Big Bang may be meaningless within standard cosmology, because "before" requires time, and time itself emerges with the universe.

More importantly for our discussion, time possesses a direction.

We experience time as flowing from past to future. The present moment becomes the past, and this progression cannot be reversed in ordinary physical processes. This one-directional behaviour is called the arrow of time.

The arrow of time is closely related to entropy.

Entropy, in thermodynamics, measures disorder or the number of microscopic configurations corresponding to a macroscopic state. In an isolated system, entropy tends to increase. This increase provides time with its direction. We remember the past because entropy was lower; we move toward the future because entropy increases.

Without entropy gradients, the distinction between past and future would lose physical meaning.

However, the detailed relationship between time, entropy, and cosmology is deep enough to require its own careful treatment. We will explore it fully in Chapter 2, where the concept that "Time is Ingenious" will be examined in detail.

For now, the essential point is this: in our observable universe, spatial dimensions and time are not independent in practice. Any real object exists within spacetime. Even if we speak of a zero-dimensional point mathematically, physically, it would persist across time. The moment we consider temporal existence, the dimensional count increases.

Thus, in strict physical reality after the Big Bang, nothing exists in purely zero dimensions. The minimal description of any physical event requires at least one temporal parameter in addition to spatial specification.

Let us now move forward.

Imagine a line.

A line has length but no breadth or height. Like the point, a perfectly one-dimensional object does not exist physically. Real objects always possess some thickness, however small. Yet the concept of a line remains foundational. It is the building block of geometry.

Extend the idea further.

A square has length and breadth. A cube has length, breadth, and height.

If we momentarily set aside the temporal dimension for simplicity, we see a pattern emerging:

- Point: 0D
- Line: 1D
- Square: 2D
- Cube: 3D

Among these, only the cube corresponds to the minimum spatial structure capable of containing volume in our universe. The line and square exist as idealisations or approximations, but not as perfectly isolated realities.

This reveals something important.

Independent dimensions combine to create richer structures. Each added dimension does not destroy the previous one. It expands it. A cube contains squares. A square contains lines. A line contains points.

Structural richness increases with dimensional addition.

Now consider this progression visually.

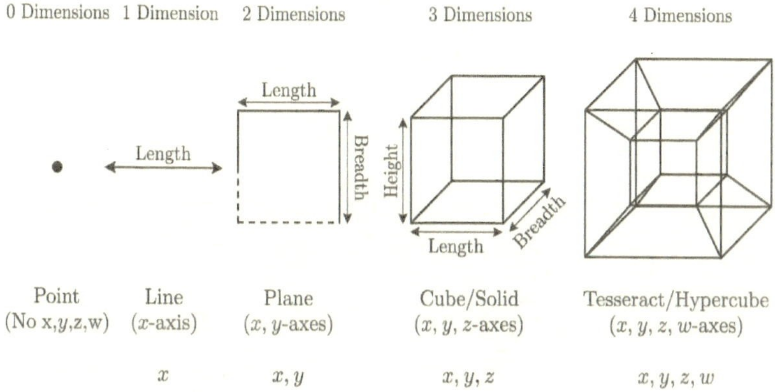

Figure 1.2: Dimensional Progression

The progression does not logically terminate at three.

We are familiar with axes labelled x, y, and z. These represent the three spatial directions. To represent a fourth spatial dimension mathematically, we introduce another independent axis, typically labelled w.

A four-dimensional hypercube, known as a tesseract, is constructed by extending a cube along this fourth perpendicular direction.

This does not end at four.

A five-dimensional hypercube, sometimes called a penteract, contains:

- 32 vertices
- 80 edges
- 80 square faces
- 40 cubic cells
- 10 tesseracts

The pattern generalises.

An n-dimensional hypercube has:

- 2^n vertices
- $n \times 2^{n-1}$ edges

This means that a 78-dimensional hypercube called 'heptaconta-octeract', *if we continue with the Greek naming style,* purely mathematical in nature, would contain:

- 2^{78} vertices
- 78×2^{77} edges

The numbers become astronomically large. Such an object has no confirmed physical existence. It belongs to mathematics, not observed physics.

But the crucial insight is this: mathematics does not impose a hard limit at three dimensions.

The limitation appears in our perception, not in logical structure.

If dimensional richness increases with added degrees of freedom, and if higher dimensions are mathematically consistent, then the possibility of existence beyond our accessible dimensional framework cannot be dismissed outright.

We have moved from a point to a hypercube.

The next question is no longer about geometry alone.

It is about hierarchy.

If structural superiority increases with dimensional access, then what does that imply about beings constrained to three spatial dimensions and one temporal dimension?

And what might exist beyond that constraint?

If One More Direction Exists

Up to this point, our discussion has remained mathematical. We have counted dimensions. We have extended cubes into hypercubes. We have written formulas for vertices and edges.

But numbers alone do not disturb intuition.

Implication does.

Let us return to a simpler comparison.

Imagine a two-dimensional world, perfectly flat. Its inhabitants perceive only length and breadth. They build enclosed shapes such as squares and circles. To them, a closed boundary is absolute. Nothing can enter without crossing an edge.

Now imagine a three-dimensional being interacting with that world.

That being could lift an object out of a sealed square without breaking its boundary. It could observe the interior of every enclosed region instantly. It could appear inside a locked room without opening a door.

From the perspective of the two-dimensional inhabitants, such actions would seem impossible.

From the perspective of the three-dimensional being, they are trivial consequences of an extra degree of freedom.

Now consider our world.

We exist in three spatial dimensions and one temporal dimension. We move freely within volume. We observe surfaces. We cannot perceive inside opaque objects without tools. We cannot bypass closed three-dimensional boundaries without physically interacting with them.

But if a fourth spatial dimension exists, the relationship between that higher-dimensional entity and us would mirror the relationship between us and the two-dimensional plane.

A being with access to an additional spatial direction could, in principle:

- Observe the interior of closed three-dimensional structures without opening them.
- Remove objects from sealed containers without breaking their walls.
- Appear within enclosed spaces without crossing visible boundaries.

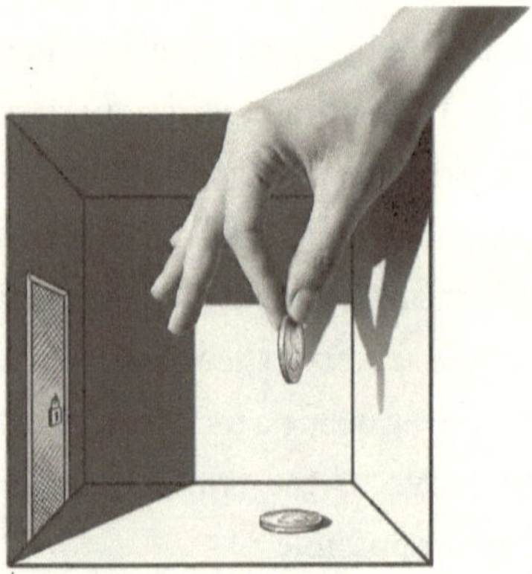

Figure 1.3: What a 3D being looks like to 2D creatures.

None of this would violate geometry from its perspective.

The limitation would be ours. The reference can be taken from *Figure 1.3*

We must be careful here. This is not evidence that such beings exist. It is a demonstration that higher-dimensional access implies structural advantages that would appear extraordinary to lower-dimensional observers.

Now extend the reasoning further.

If the progression from zero to three dimensions has increased structural richness at every step, and if theoretical physics allows more than four total dimensions, then the hierarchy does not end where our perception ends.

The important realisation is this:

Dimensional superiority is not about power in a mystical sense. It is about degrees of freedom.

An entity with more independent degrees of freedom has access to a larger configuration space of reality.

We operate within three large spatial dimensions and one temporal dimension. String Theory proposes additional compactified spatial dimensions. M-Theory extends this further.

Even if those extra dimensions are curled up at extremely small scales, their existence means the total structure of reality is larger than what we directly experience.

And if the structure is larger, then so is the possibility.

The next step is no longer geometric imagination.

It is theoretical physics.

If higher dimensions are not merely mathematical constructs but arise from consistency requirements of physical theories, then we must examine what those theories actually say.

This is where we move from counting axes to confronting M-Theory itself.

And only then can we responsibly return to the original question:

If dimensional hierarchy is real, where do we stand within it?

Rendering Possibilities

You are turning the pages, probably reading them too, for the answer to the query raised at the beginning of the chapter -*"Who - Where & Why ?"*

We now finally begin, *'The Rendering of Possibilities',* where we will thoroughly analyse each dimension, its suitability for 'The Watcher', and what access it can provide, being that 'x-dimensional' creature.

It is important to recognise that the qualities we attribute to each dimension are not fixed or automatically sequential. Dimensions represent independent degrees of freedom, which are structural in nature, but they do not inherently grant specific abilities.

A two-dimensional world, for example, could be defined by length and breadth, or equally by length and height; the choice of axes is arbitrary, yet each defines a plane of existence. Just as a flat creature could never perceive the third direction, a higher-dimensional being may access certain freedoms, but how those freedoms manifest depends on context, interaction, and the underlying physical laws.

The progression from zero to eleven dimensions is a conceptual scaffold, a tool to visualise increasing structural richness and explore what might be required for the Watcher's perspective. Some capabilities we might associate with a higher dimension could already exist in a lower one if the axes are combined differently, while others only emerge through the interplay of multiple dimensions.

The numerical order we use is illustrative, a guide to comprehension rather than a strict hierarchy of power or awareness. We begin at the most abstract beginning, a point. Zero dimensions. A location without extension. No length. No breadth. No height. It is pure potential, pure abstraction. In the physical world, it cannot exist independently, but as a conceptual building block, it is invaluable. A zero-dimensional entity could not influence space. It could not act, move, or perceive. It is entirely confined to the mathematical realm. Even the simplest notion of "control" is meaningless here. If the Watcher were zero-dimensional, creation itself would be impossible. Such a being could never shape a line, never form a plane, never induce motion. The very concept of a universe, with events and consequences, would lie beyond its grasp.

From zero, we step to one dimension—a line. Length alone defines existence. A one-dimensional being can move forward or backwards, but only along a single axis. It can exert influence only in that linear direction. It could stretch across a distance, perhaps intersect another line, but interaction is trivial, constrained. A being of this nature might manipulate trajectories along that axis, but it could not encompass space as we know it. If we attempt to imagine such a creature as the Watcher, we immediately see the impossibility. To create a cosmos of complexity, a single line is not enough. A line is directional, constrained, incapable of comprehending the breadth of a two-dimensional plane, let alone the volume of three-dimensional space.

Two dimensions expand the canvas. A plane has length and breadth. A two-dimensional being can move and act across this plane. It can observe relationships in two directions simultaneously. Imagine, as before, the flat civilisation: squares, triangles, circles. A two-dimensional observer can fully comprehend all shapes on the plane. Yet, even here, there are limits. Height is inaccessible. What lies above is invisible. Such a being could manipulate areas, redirect flows within its plane, but it cannot reach into the third dimension. To a three-dimensional observer, its actions appear limited. The two-dimensional entity may attempt to create, but all influence is confined. It lacks access to depth, to volume, to the inner structure of objects. For the Watcher, whose cognition must encompass the universe, a plane is insufficient. Complexity cannot arise within flatness alone.

Three dimensions—the world we know. Length, breadth, and height.

Now, a being can navigate volume, perceive surfaces, and interact with objects in a familiar space. Creation becomes possible, but only within these constraints. A cube can be formed. Matter can be shaped. Forces can act. Yet, even here, limitations remain. Observation is line-of-sight. Internal structures may remain hidden. Closed systems require entry or intervention. Though potent relative to lower dimensions, the three-dimensional being is still limited. The universe, in its full richness, extends beyond any single plane or surface. To be the Watcher, one must transcend even this cubic cage.

Four dimensions introduce time. Length, breadth, height, and temporal flow. Now events are not only located but sequenced. Cause and effect are visible. A four-dimensional being could, in principle, move along the time axis as freely as the spatial axes. The past, present, and future become simultaneously accessible. This implies tremendous control over processes, an ability to observe outcomes before they manifest. Yet temporal navigation is not equivalent to omniscience. Awareness over time still does not guarantee comprehension of all possibilities. A four-dimensional being might see the future but cannot necessarily alter every contingent variable outside its immediate axis. Still, we approach a scope worthy of a superior entity. A temporal dimension provides narrative, but not the infinite control we associate with the Watcher.

Five dimensions open a space of alternative possibilities. In addition to the familiar spacetime, a fifth axis allows branching outcomes—different choices, different histories. Here, influence is no longer linear. A being with access to five dimensions can perceive multiple potential futures simultaneously. It can explore paths not taken and outcomes not realised. However, branching alone does not imply judgment or purpose. A five-dimensional entity can observe countless scenarios, but without higher-order integration, it may lack the discernment to manage them. Structural access expands, but cognition does not automatically follow. The Watcher must integrate awareness with agency, not merely potentiality.

Six dimensions add further layers of causality and variation. Now multiple timelines, universes, and probabilistic outcomes exist concurrently. Influence extends across broader configuration spaces. A six-dimensional being could, in principle, understand interactions across parallel realities, anticipate consequences of causal chains that span universes. Yet again, comprehension is not guaranteed. Complexity multiplies. The ability to act coherently across this expanded framework requires extraordinary structuring of cognition. Control without understanding is chaos. Even here, the Watcher's suitability cannot be assumed.

Seven dimensions introduce meta-level conditions—spaces of rules themselves. Here, a being could potentially perceive not just events, but the frameworks within which events arise. It could discern laws governing each universe and observe how constants and structures vary. Influence expands to the architecture of possibility itself. Yet even this is partial. To truly be the Watcher, one must encompass not only rules but meta-rules, the very scaffolding of reality in full. Seven dimensions hint at the architecture, but the apex has not yet been reached.

Eight dimensions increase abstraction. Probabilities, outcomes, and meta-rules intertwine. Influence can extend to higher-order structures. Yet complexity becomes formidable. Each additional dimension multiplies possibilities exponentially. Awareness is taxed; cognition must be correspondingly vast. Structural superiority grows, but without integrated understanding, omniscience remains elusive.

Nine dimensions approach to the threshold proposed by string theories. Here, all spatial and temporal possibilities intertwine with hidden dimensions that influence physics subtly yet profoundly. A nine-dimensional being could, in principle, see the mechanics of creation, understand forces at scales inaccessible to lower-dimensional observers, and integrate knowledge across realities. But perception is still constrained by the inherent limits of cognition. To truly be the Watcher, one must unify access with comprehension.

Ten dimensions, as described by superstring theory, encompass every spatial and temporal degree of freedom plus compactified dimensions curled at minuscule scales. These influence particle behaviour, forces, and constants of nature. A ten-dimensional entity can manipulate physical laws in its immediate universe and anticipate outcomes with remarkable fidelity. It approaches the structure needed for ultimate creation, yet the hidden dimensions imply subtlety and intricacy that must be mastered. Without mastery, even ten-dimensional existence is insufficient for full cosmic oversight.

Finally, eleven dimensions—the apex suggested by M-Theory. Nine spatial dimensions plus one temporal dimension, with additional compactified structures. Here lies the theoretical framework sufficient for a being capable of overarching creation, the Watcher. An eleven-dimensional entity possesses structural access to all known forces, can influence outcomes across a multitude of realities, and comprehends the hidden scaffolding of matter itself. Cognition and structural access converge. This is the minimal framework that allows for the creation of universes, the orchestration of probabilities, and the implementation of laws.

Yet a caution remains. Structural access, even at eleven dimensions, does not automatically produce perfect awareness or judgment. Consciousness, insight, purpose—all remain emergent qualities. The Watcher, we infer, is not merely a collection of axes or dimensions. It is the integration of structure, comprehension, and will. It must be able to navigate the immense configuration space, discern coherence among countless possibilities, and act decisively. Here, finally, the conceptual framework allows for a being whose existence matches the title: The Watcher.

In this segment, we have traversed the ladder from zero to eleven dimensions. We have examined what control each dimensional being could exercise and evaluated their suitability to be the Watcher. Only at the apex, within eleven dimensions, does the structure provide the possibility for a being capable of creation, observation, and governance of reality itself. But even this realisation is tempered by the reminder: structural potential is necessary but not sufficient. Consciousness, integration, and discernment are essential complements to dimensional supremacy.

The Infinite Watcher

At this point, it may be tempting to conclude that the Watcher, the architect of all existence, must inhabit eleven dimensions, just as M-Theory suggests for the consistency of physical reality. After all, if additional dimensions grant increasing structural richness and control, would the ultimate creator not reside in the highest-known dimensional framework? The answer, however, is far from simple.

To begin with, it is necessary to remind ourselves of the limits of human knowledge. In the early 20th century, physics recognised only three spatial dimensions. Time was considered a separate entity, a universal clock ticking independently of space. Later, the understanding of spacetime by Albert Einstein revealed that space and time are interwoven into a four-dimensional continuum. Still later, the development of String Theory and M-Theory suggested that additional compactified dimensions could exist, hidden from our direct observation yet necessary for theoretical consistency. Today, eleven dimensions are considered by many physicists as a framework for unifying the known forces of nature. But these numbers are provisional. They are not absolute truths about the universe—they are models that extend the mathematics we can conceive and the observations we can make. There is no reason to assume the universe cannot contain far more dimensions than we currently know. In this sense, even the notion of "the highest dimension" is a limitation imposed by our understanding, not by nature itself.

Furthermore, structural access through dimensions does not automatically confer cognitive superiority. Dimensions provide degrees of freedom, directions along which existence can extend and unfold. They offer potential for control, observation, and interaction with reality. But the ability to inhabit higher-dimensional space does not necessarily equate to omniscience or omnipotence. A being existing in twelve dimensions may possess extraordinary structural freedom, yet without awareness and understanding, it remains no closer to the role of a creator than a three-dimensional human restricted to the familiar x, y, and z axes. Consciousness is not a byproduct of dimensionality—it is an emergent quality that may or may not accompany higher-dimensional existence.

Entropy adds another layer of complexity. Within our observable universe, time flows in a single direction—the arrow of time—from past to future. Entropy, the measure of disorder, provides this arrowhead. Every dimension we observe is entropic; all processes move toward increasing entropy. For a being that creates or governs such a universe, the act of creation inevitably interacts with entropy. To originate a universe with time flowing and entropy increasing is, paradoxically, to participate in the constraints that time imposes. And yet, this does not define the creator entirely. Perhaps some universes are non-entropical, devoid of a directional flow of time, existing instead in a chaotic or timeless state. The ultimate Watcher must be capable of perceiving, understanding, and existing in both entropical and non-entropical realities simultaneously. Its cognition cannot be limited by the structure of any single universe or the direction of entropy.

We must also acknowledge that the very notion of a finite number of dimensions is limiting. Eleven, ten, or even a hundred dimensions are concepts derived from our mathematical models. They are tools for calculation, for predicting particle behaviour, or for constructing consistent theoretical frameworks. But they do not define reality itself. To assign the Watcher a specific dimensional number is to impose human constraints on what is, by necessity, unconstrained. The Watcher is not a resident of eleven dimensions; it is beyond the countable, the enumerable, and the observable. To imagine it as "eleven-dimensional" is to misunderstand the nature of infinity itself.

Another critical consideration lies in the interplay between dimensions and cognition. Each additional dimension expands the configuration space available to a being—it increases potential positions, interactions, and structural relationships. Yet the capacity to process, perceive, and manipulate these degrees of freedom is not guaranteed. A higher-dimensional being may possess the ability to occupy more space, to interact with more variables, or to access more pathways within the structure of reality, but without consciousness, these abilities are meaningless. The Watcher, as the creator, must not only exist within—or beyond—these dimensions but must also wield comprehension across all of them. Its awareness is not constrained by human measures of intelligence; it is inherently tied to the totality of structural possibility.

To illustrate, we may return to the analogy of lower-dimensional worlds. A creature confined to a flat, two-dimensional plane perceives only length and breadth. It can understand areas and perimeters, enclose spaces with squares or circles, and operate entirely within its plane. A three-dimensional being interacting with this plane can manipulate objects in ways incomprehensible to the two-dimensional inhabitants. It can reach inside sealed regions, lift objects without crossing boundaries, and observe interiors without breaking surfaces. Structural freedom alone gives the higher-dimensional being the appearance of superiority—but it is the combination of freedom and comprehension that defines true mastery. The Watcher, then, is not merely a being of many dimensions; it is a being whose cognition and awareness fully exploit the infinitude of dimensions it inhabits.

In contemplating the Watcher, we confront the unreasonable. Every framework, every model, every measure is provisional. We may count dimensions, extend cubes into hypercubes, or visualise eleven-dimensional space—but the creator of all is beyond these constructs. It is infinite-dimensional, omnipresent in the possibilities of reality, and unbound by the limitations that define our observable universe. To glimpse the Watcher is to recognise the vastness beyond the measurable, the eternal beyond time, and the omnipotent beyond structure.

Denouement I

You might be thinking, "I came all this way not just to receive a mysterious answer, or one that itself seems uncertain." And in a sense, that expectation is justified. But the answer, while subtle, can be given—and understood—through the lens of everything we have explored so far.

After exploring the scaffold of reality, counting dimensions, and imagining beings in spaces beyond our perception, we arrive at the inevitable question once more: where does the Watcher reside?

Throughout this journey, we have moved carefully, step by step—from a single point, to a line, to a cube, to hypercubes, and finally to the abstract spaces described by M-Theory. We have examined the relationship between structural freedom and potential awareness, the constraints imposed by entropy and time, and the implications of multiple, perhaps infinite, universes. At every stage, the pattern is clear: the addition of dimensions increases the degrees of freedom, allows greater control over reality, and expands structural possibilities. But it does not automatically confer omniscience.

Thus, the final verdict emerges: the Watcher is not constrained by eleven dimensions, nor by any countable framework we can conceive. To assign a number is to impose human limits on an infinite concept. The Watcher exists in the full spectrum of structural possibility—every direction, every degree of freedom, every dimension, known and unknown. Its domain is the totality of reality, not just the segment we can describe mathematically or observe physically.

In other words, the Watcher is infinite-dimensional, beyond the confines of the observable universe, beyond the scope of any mathematical model, beyond the limitations of entropy and temporal flow. To seek the Watcher in a particular dimension is to misunderstand the nature of creation itself. The Watcher is not "found" as a being in a room or a coordinate in space; it is the totality of possibility made conscious, the ultimate framework in which existence itself unfolds.

At last, we have an answer, a little absurd but an answer. Though you might think that the answer could've been just a sentence that 'The Watcher' is an infinite-dimensional being, yes, you are correct, but there is a catch. This statement, though absurd, would mean nothing if all the other dimensions were not considered. There are still some possibilities that we can consider to continue our search outside 'Our Observable Universe.'

And so the next question arises: if dimensions can stretch toward infinity, what becomes of time itself? Is it a fixed arrow, or merely a perception shaped by our limited view?

2

Time Is Ingenious

" 'Tis not a gap, but a void that is stretched indefinitely."

We wake up to clocks, live by calendars, and measure our existence in seconds, minutes, and years. Everything around us seems to move in one direction, from past to present to future.

A broken glass never reassembles itself. A child never grows younger. The sun rises, sets, and rises again, as if obeying an invisible command. From this everyday rhythm, we assume that time must be real, steady, and absolute. But what if this steady flow is not a property of the universe, but a construction of the mind trying to organise change? So, you might ask -
Is time a fixed arrow, or merely a perception by our limited view?

If time truly flows, where exactly is it flowing? Is it embedded in the fabric of the cosmos, or does it emerge from motion, memory, and entropy? When we say something happened before or will happen after, are we describing an external dimension or simply arranging experiences in a sequence our brains can understand? Perhaps what we call time is not a universal river carrying everything forward, but a shadow cast by deeper physical laws we have yet to grasp fully. To somewhat understand what time is? We need to first know something which the time obeys - Entropy.

Entropy - The King of Thermodynamics

Entropy is often introduced in textbooks as a measure of disorder. While that description is not wrong, it is incomplete. Disorder is only a surface-level picture. Entropy is more accurately a measure of how many possible microscopic arrangements can produce the same visible state of a system. In simple words, it tells us how many hidden ways something can exist while still appearing the same to us.

Imagine a box divided into two halves by a thin partition. On one side, all the gas molecules are trapped. On the other side, there is nothing. This arrangement looks neat and organised. The gas is confined. This can be seen in *Figure 2.1,* where the left side represents a low entropy state and the right side, a high entropy state.

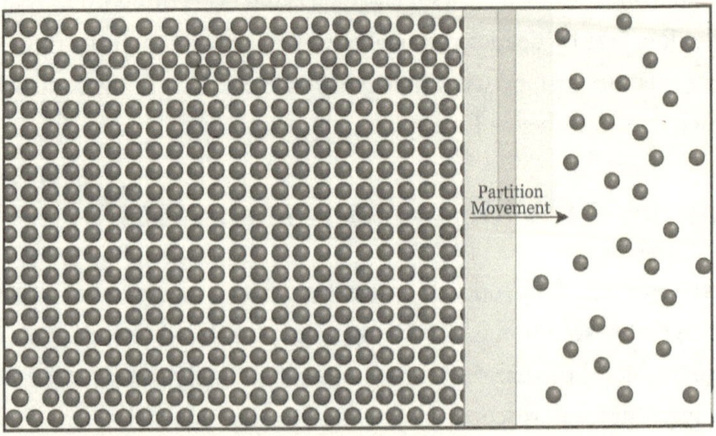

Figure 2.1: Entropy Difference in a Box.

The moment the partition is removed, the molecules rush and spread across the entire box. Soon, they fill every corner. Now compare the two situations. In the first case, there is only one very specific arrangement where all molecules stay in one half. In the second case, there are countless arrangements in which the molecules can be spread out. The second situation has far more possible microscopic combinations. That is why we say it has higher entropy.

Entropy, therefore, is not about visible messiness alone. It is about probability. The universe naturally moves toward states that are statistically more likely. And states with higher entropy are overwhelmingly more likely than states with lower entropy.

This idea was first formally introduced by the physicist Rudolf Clausius in the nineteenth century. He studied heat engines and asked a profound question. Why can heat flow from a hot object to a cold one, but never spontaneously the other way around? Why does a cup of hot tea cool down in a room, yet the room never cools down by giving heat back to the tea? Clausius realised that there was a direction embedded in nature itself. That direction was described by what we now call the Second Law of Thermodynamics.

The Second Law states that in an isolated system, entropy never decreases. It either increases or remains constant. This simple statement has enormous consequences. It explains why certain processes are irreversible. A broken egg does not reassemble itself. Smoke does not gather back into a candle. A shuffled deck of cards does not naturally return to perfect order. These are not forbidden by the laws of motion. They are forbidden by probability.

Later, Ludwig Boltzmann gave entropy a deeper meaning. He connected entropy with statistics and microscopic motion. He proposed a famous relation which, in words, tells us that entropy increases when the number of possible microscopic arrangements increases. On his grave is engraved the formula that links entropy with probability. That equation changed physics forever. It revealed that thermodynamics is not separate from mechanics. It is mechanics viewed through the lens of large numbers.

To understand this better, consider a simple analogy. Suppose you toss a thousand coins. There are many more ways to get roughly five hundred heads and five hundred tails than to get all heads. The balanced outcome is statistically favoured, not because nature prefers balance morally, but because it is overwhelmingly more probable. Similarly, a system naturally drifts toward the state that can occur in the greatest number of ways.

Entropy is therefore deeply connected to information. If a system has low entropy, we have more information about its exact configuration. If all gas molecules are in one half of the box, we know something very specific about their positions. If they are spread randomly everywhere, our information about any single molecule becomes limited. The system becomes more uncertain at the microscopic level.

In the twentieth century, Claude Shannon discovered that the mathematics of entropy also applies to communication and information theory. Shannon entropy measures uncertainty in messages. The same logic that describes heat and molecules also describes bits and data. This was not a coincidence. It suggested that entropy may be a fundamental bridge between physics and information.

But we must go deeper still.

Entropy is not simply chaos. It is the count of possibilities. A tidy room is a very special arrangement of objects. There are countless ways to scatter clothes and books randomly. That is why rooms become messy easily and require effort to stay clean. Effort means energy input. Maintaining low entropy requires work.

This reveals another powerful truth. Local entropy can decrease, but only at the cost of increasing entropy somewhere else. A refrigerator cools its interior, creating order inside. But it releases heat into the surrounding room, increasing overall entropy. Life itself is an example of this principle. Living organisms maintain complex structures and organisation. They build cells, tissues, and brains. But they do this by consuming energy and releasing heat into their environment. Globally, entropy still increases.

So entropy is not the enemy of order. It is the price of existence.

Now, let us confront a subtle but essential point. At the microscopic level, the fundamental equations of physics do not prefer a direction of time. If you record two particles colliding and play the video backwards, the reversed motion still obeys Newton's laws. The same holds in quantum mechanics. The equations do not care about the past or future. Yet in our world, processes clearly move in one direction. We remember yesterday, not tomorrow. We grow older, not younger. Ice melts in warm water, not the reverse.

This direction arises because of entropy.

The increase of entropy gives time its arrow. Without entropy, the distinction between past and future would blur. The reason we remember the past is that memories are physical structures in our brains. Creating a memory increases entropy. The past is fixed because it corresponds to lower entropy configurations that have already evolved into higher entropy states. The future is open because it corresponds to many possible higher entropy arrangements.

In this sense, entropy is not just a thermodynamic quantity. It is the foundation of temporal experience.

Yet, entropy does not mean everything is descending into chaos instantly. The increase is gradual. It follows statistical laws. Systems move from less probable states to more probable ones over time. Sometimes fluctuations occur. In very small systems, entropy can momentarily decrease. But on large scales, the overall trend dominates.

Another misunderstanding is to think of entropy as purely destructive. Entropy indeed leads to decay. Stars exhaust their fuel. Structures erode. Energy spreads out. But entropy also enables complexity. The formation of galaxies, stars, and planets occurred because the early universe was in a very special, low-entropy state. As entropy increases, energy flows from concentrated regions to emptier ones. These flows create structure temporarily. Without entropy gradients, nothing dynamic would occur.

Think of a waterfall. Water flows because there is a height difference. Once everything is level, the flow stops. Similarly, energy flows because of entropy differences. When equilibrium is reached, no more useful work can be extracted. Motion becomes meaningless.

When I'm declaring entropy as the *'King of Thermodynamics',* I'm not simply exaggerating it poetically. Every engine, every star, every living organism obeys its rule. The Second Law has never been observed to fail. It is one of the most reliable principles in science.

And yet, it raises deeper questions. If entropy always increases, how did the universe begin in such a low entropy state? If entropy defines the arrow of time, what does that say about time before the beginning? Can entropy ever reach a maximum? Can it stop?

The Entropy's Ruling

To ask how entropy started is almost the same as asking how the arrow of time began. Because without entropy, there is no direction. There is no before and after in any meaningful sense. So to understand the birth of entropy, we must travel back to the birth of the universe itself.

According to modern cosmology, the universe began approximately 13.8 billion years ago in what we call the Big Bang. This was not an explosion in space. It was the rapid expansion of space itself from an extremely hot, dense state. In its earliest fraction of a second, the universe was unimaginably small, energetic, and uniform.

Now here lies the great puzzle.

If entropy always increases, then the universe must have started in a state of very low entropy. Otherwise, there would be no room for it to rise. But how could something so hot and energetic be low in entropy? Does not heat mean disorder?

The answer is subtle.

In the early universe, matter and radiation were spread almost perfectly evenly. At first glance, this seems like maximum disorder. Everything was mixed uniformly. But gravity changes the story. When gravity is taken into account, a uniform distribution of matter is actually a highly ordered state. Why? Because gravity prefers clumping. Given enough time, matter pulls together to form stars, galaxies, and black holes. A perfectly smooth universe is statistically rare when gravity dominates.

This insight was strongly emphasised by physicist Roger Penrose. He argued that the early universe had extraordinarily low gravitational entropy. The smoothness of the cosmic background radiation shows us that matter was distributed with incredible precision. That smooth beginning is what allowed entropy to increase over billions of years through gravitational clumping.

In other words, entropy did not begin in chaos. It began in astonishing order.

As the universe expanded, tiny fluctuations in density grew under gravity. Slightly denser regions attracted more matter. Over time, these regions formed stars and galaxies. Nuclear fusion ignited inside stars. Light flooded the space. Heavy elements were forged. Planets formed. Eventually, life emerged on at least one small world.

All of this complexity arose because entropy was increasing.

It may sound contradictory, but the growth of structure does not violate the Second Law. When a star forms, matter becomes more ordered locally. But the process releases enormous amounts of radiation into space. That radiation increases the total entropy far more than the decrease caused by clumping. The universe as a whole still moves toward higher entropy.

So entropy began its rise the moment the universe began evolving from that special low entropy state.

But this pushes the mystery one step deeper. Why was the universe in such a low-entropy condition to begin with? However, no theory has yet provided a fully satisfying explanation for why the entropy of the early universe was so extraordinarily low. It remains one of the deepest unsolved problems in physics. If you ask my opinion, I would say, *"Someone or Something, call it The Watcher or not, created it in a very precise manner and left space for entropy to increase."* Finding an answer to this is like proving God with science. There are also cyclic universe theories suggesting that what we call the beginning might be a transition from a previous cosmic phase.

Whether the universe began from a singularity, a bounce, or a quantum event, the crucial fact remains that it began in a low-entropy configuration.

And from that moment onward, entropy began its ascent. As space expanded, temperature dropped. Energy gradients formed. These gradients allowed work to be done. Stars burned their fuel. Chemical reactions occurred. Biological systems harnessed energy differences. Everywhere, entropy increased overall, even as local pockets of order temporarily flourished.

It is important to realise that entropy did not start ticking like a clock at a specific second. It is not a substance that was switched on. Entropy is a property of arrangements. The moment there were distinguishable states and probabilistic configurations, entropy became meaningful. In that sense, entropy began when the universe became describable in terms of physical states. Before that, the concept may not even apply.

If the early universe had begun in maximum entropy, nothing interesting would have followed. No stars, no galaxies, no chemistry, no life. Everything would already be in equilibrium. The reason anything happens at all is that entropy had room to grow.

So when we ask how entropy started, we discover something remarkable.

When Does Entropy Die?

If entropy has been increasing since the birth of the universe, the natural question arises: Will it ever stop? And if it does not stop, what does that mean for everything that exists?

The honest scientific answer is this. According to everything we currently understand, entropy will not stop increasing as long as the universe continues to exist in its present physical framework. The Second Law of Thermodynamics is statistical, not absolute in a rigid mathematical sense, but for systems as large as the universe, statistical certainty becomes indistinguishable from inevitability.

However, there is a difference between entropy increasing and entropy having somewhere left to increase. Entropy does not grow infinitely fast. It increases toward a maximum possible value, a state known as thermodynamic equilibrium. In that condition, energy is spread out as uniformly as possible. There are no gradients, no temperature differences, no concentrated energy pockets to perform work. When that state is reached, large-scale change effectively ceases.

This predicted cosmic future is often called the heat death of the universe. It does not mean flames everywhere. It means stillness. Stars will exhaust their nuclear fuel. Galaxies will drift apart as space expands. Black holes will eventually evaporate through quantum effects described by Stephen Hawking. Matter may decay. Over unimaginable spans of time, the universe would approach a thin, cold distribution of energy where no organised processes can occur.

In such a state, entropy has not stopped increasing because it was defeated. It has stopped because it has reached its ceiling. There is no further statistical direction to move toward.

Having vividly understood entropy. Now, a deeper question might arise in your head: But is there anything that can defy entropy?

The answer will be both a 'yes' and 'no.' As there are conditions to it-

Entropy governs every physical process involving energy exchange. It dictates how engines operate, why fuels burn, and why no machine can be one hundred percent efficient. A perpetual motion machine of the second kind is impossible because it would require entropy to decrease without compensation.

On a cosmic scale, entropy shapes the life cycle of stars. A star shines because nuclear fusion converts mass into energy, increasing entropy overall. When fuel runs out, the star collapses or disperses. Even black holes, once thought to be perfect absorbents that trap everything forever, possess entropy. In fact, black holes may have the highest entropy of any object for their size. Their entropy is proportional to the area of their event horizon, a discovery linked to the work of Jacob Bekenstein and later refined through Hawking's calculations.

At first glance, living organisms appear to defy entropy. They grow, repair themselves, and maintain extraordinary internal order. Cells organise molecules with precision. DNA encodes structured information. The human brain stores memory and thought. It seems as if life pushes against disorder. Yet life does not escape entropy. It negotiates with it.

A living organism is not an isolated system. It constantly exchanges energy and matter with its surroundings. Consuming food or sunlight, it draws in low-entropy energy. It then releases higher entropy waste heat into the environment. The organism maintains internal structure by increasing entropy externally. The Second Law remains unbroken.

Is there anything that can escape entropy completely?

If something exists physically within our universe, it must obey the statistical structure of the universe. Even information, once thought abstract and separate, has a thermodynamic cost. Erasing a bit of information requires energy and increases entropy, a principle known as Landauer's principle.

One might imagine a hypothetical entity beyond spacetime, untouched by physical processes. Such a being, if it existed, would not be bound by thermodynamics because thermodynamics describes physical systems. But as long as we remain within observable physics, no material object, no star, no organism, no machine can stand outside entropy's influence.

Entropy ensures that moments differ from one another. It creates the possibility of evolution, growth, and experience. It also guarantees decay and eventual stillness.

Now that we know about a major part of entropy, we can begin with the comprehension of the integrants of time.

Hourglass

You might have heard, *"Entropy is the arrowhead of time"*, but why do we say that? Because entropy gives time its direction.

A shattered glass does not leap from the floor and assemble itself. Heat flows from hot to cold, not cold to hot. We grow older, not younger. The reason for this asymmetry is not hidden in the equations of motion. It lies in entropy.

Entropy increases toward states that are statistically more likely. When a glass falls and breaks, the fragments can arrange themselves in an enormous number of microscopic configurations. The intact glass corresponds to a very specific and rare arrangement of molecules. The broken pieces correspond to vastly more possible arrangements. The universe naturally evolves toward the overwhelmingly probable state. This statistical direction is what we perceive as the forward flow of time.

The arrow of time is not imposed from outside. It emerges from the fact that the universe began in a low-entropy state. From that special beginning, entropy has been increasing. Because entropy increases, processes become irreversible. Because processes are irreversible, we distinguish past from future.

But how exactly is the direction determined?

Imagine a universe that begins in maximum entropy. Everything is already in equilibrium. Energy is evenly distributed. No gradients exist. In such a universe, no large-scale change would occur. Without change, there would be no distinguishable sequence of events. Time, as a meaningful concept, would lose relevance. Direction emerges only when there is a gradient.

The early universe was extraordinarily low in entropy, especially in gravitational terms. That low entropy condition created the possibility for entropy to rise. As entropy increases, the universe evolves through distinct states. The direction from lower entropy to higher entropy defines what we call the future.

So the arrow of time is not absolute. It depends on boundary conditions. If the universe had begun in a different state, our perception of time might be entirely different.

At this point, an even deeper question can be asked: What would happen to time if entropy stopped increasing?

If entropy reached its maximum possible value, the universe would enter thermodynamic equilibrium. No temperature differences. No energy gradients. No usable energy. All processes that depend on imbalance would cease.

In that state, change would effectively stop.

Without change, measuring time becomes meaningless. Time is defined through events. A clock ticks because of periodic motion. Motion requires energy differences. If all differences vanish, nothing oscillates, nothing decays, nothing transforms. Time would not vanish mathematically, but it would lose operational meaning. It would become an empty parameter without events to mark its passage.

The hourglass flows because gravity pulls sand downward. The cosmic hourglass flows because entropy increases.

Time Dilation

I call time ingenious because we perceive time in a very clever manner. One of the reasons is that entropy gives time its direction, gravity and motion shape its rate. Time does not pass at the same speed everywhere. This is one of the most astonishing discoveries of modern physics. According to the theory of relativity, time can stretch and compress depending on motion and gravity.

There are two primary forms of time dilation. Let us consider them both:

The first arises from motion. If an object moves at a speed close to the speed of light, time for that object slows down relative to a stationary observer. This effect is called special relativistic time dilation. Astronauts travelling at extremely high speeds would age slightly less than people on Earth. The faster you move through space, the slower you move through time.

The second arises from gravity. Massive objects curve spacetime. The stronger the gravitational field, the slower time passes relative to regions with weaker gravity. This is gravitational time dilation.

For example, near a black hole, time slows dramatically. To a distant observer, a clock near the event horizon appears to tick slower and slower. In extreme cases, it appears almost frozen. This is not an illusion. It is a consequence of spacetime curvature.

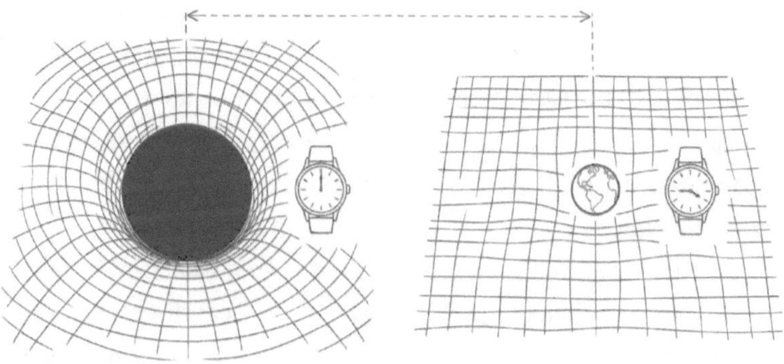

Figure 2.2: Time Dilation Because of Spacetime Curvature.

Time is not a universal metronome ticking uniformly across the cosmos. It is woven into the geometry of spacetime. Where gravity is strong, spacetime bends deeply, and time stretches. So why does it occur? There is something which we won't discuss in depth here, which is that the speed of light must be constant. It is not relative. Now, according to relativity, spacetime can be compared to a mesh-like fabric. When objects like planets or even bigger celestial bodies with an immense gravitational field are placed on the mesh, they create a curvature, and that is what we call gravity.

When light is travelling near these massive bodies, it bends along the spacetime curvature. The situation starts to get messy at this point. Why? Because the distance between Point 'A' and Point 'B' is the same, it has not changed, but due to the curvature, light needs to cover that extra distance in the same amount of time. Which it obviously can't. It will be seen that light is taking more time to cover the same distance, changing the value of its speed, which is distance/time. To prevent this from happening, the universe dilates time. It makes the time flow slowly near objects with a gravitational field, not only near black holes, but also near every object with a gravitational field. We take the example of a black hole because the effect is prominent near such regions and very slight, not even noticeable, near stars like the Sun, or planets like Earth.

Let me tell you, we have experimented with it too. How? Well, the spaceships in which the astronauts travel it travels with a great speed, at least 11.2 Km/s which is the escape velocity of Earth. It is noticed that the watches that the astronauts wear run slightly slow; the difference is minuscule, nearly 10 nanoseconds in a year. But it is a number good enough to prove that time dilates with an increase in velocity. Not only with velocity, but with gravity too. When atomic clocks are taken to places on Earth where the gravitational force of attraction is slightly less, they are seen to run slightly faster, nearly 7 nanoseconds, but again, a number reasonable enough to prove time dilates.

Now, you know what time dilation is. But it has its astonishing implications too. Some even sound like fantasy, like time travel and particle life extension. Let's start with particle life extension, because why not keep the deserts for the meal's end? The candidate particle we have is called 'A Muon'. What is it? We'll look at it later, for now, just think that it is a particle like atoms, protons or neutrons, though it is much smaller.

Muons have an estimated lifespan of nearly 2.2 microseconds and travel at nearly 99% of the speed of light. If we do the math, they should cover the distance of approx 660 meters before it loses its existence. But to everyone's surprise, it covers about 10 - 15 kilometres. And not only that, any observer at rest will see the muon exist for nearly 10 - 20 microseconds. But wait, there's a catch. For the muon itself, its lifespan is still 2.2 microseconds; for us, it appears to be dilated. If you were a muon, you would exist for just 2.2 microseconds, but I would see you exist for 5 to 10 times more than that.

Now, the time travel. First, don't think of creating something like a grandfather paradox, as going back in time is forbidden in this universe; what has happened, it has happened, and it can't be changed. Going back in time is forbidden, so we can go in the future, right? Yes. But let me tell you, it is a one-way ticket. Once you go into the future, there is no coming back; you become a resident of that time. Want to try it? Sorry to say, but you can't. Why? Because, to do that, either of two things is required. The first option is that you can travel at extremely high speeds, say, like 90 - 99% of the speed of light and time will be so much dilated that a few seconds for you will be years on earth and for the people at rest or at extremely low speed. Why is travelling at such speed not possible? We'll discuss it later in a very vivid manner. For now, let's explore the second way. The second way to do it is to travel to a supermassive black hole. Stay near its horizon for a few minutes or hours, depending on how far you want to travel in the future and then travel back to Earth. You will notice that a few moments you spent near that giant has equated to years or decades on Earth. Again, it's not possible. Why? The nearest supermassive black hole to Earth, Sagittarius A*, is approx 26,000 light-years away. Even if you travel at the impossible speed of light, I don't think that you'll live 26,000 years. Not only this, imagine what would happen to you if you came close enough to the vicinity of that black hole. Something that engulfs stars and planets in itself; leave, don't think about it. A short note: There are several things mentioned, which I haven't really explained, for we will do it in the later chapters.

Now, coming back to entropy, let us connect time dilation to it. Even though time can dilate, the Second Law still holds locally. Entropy increases along every observer's own timeline. No matter how slowly time passes near a black hole, entropy within that local system continues to rise in its own frame.

Time dilation changes the rate at which time flows relative to another observer, but it does not reverse the arrow defined by entropy.

Cyclic Universe Models

After walking through entropy, the arrow of time, time dilation, and the possible stillness of heat death, we now arrive at a daring alternative. What if the hourglass does not empty once and remains empty forever? What if, after the last grain falls, the universe turns itself over?

Cyclic universe models propose exactly this. Instead of a single beginning followed by eternal expansion, they suggest that the universe may undergo repeated phases of birth, evolution, collapse, and rebirth. In such a framework, the cosmos is not a one-time event. It is a recurring phenomenon.

To understand these models, we must connect them carefully to what we have already established.

We learned that entropy defines the arrow of time. We saw that the universe began in an extraordinarily low entropy state. From that rare beginning, entropy has been increasing for billions of years. If nothing intervenes, the universe approaches thermodynamic equilibrium, often described as heat death. Cyclic models challenge this finality.

In the standard cosmological model, the universe began with the Big Bang and has been expanding ever since. Observations show that this expansion is accelerating, likely driven by dark energy. If this acceleration continues indefinitely, galaxies will drift apart, stars will die, and entropy will approach its maximum.

However, the equations of general relativity allow more than one possibility. They permit a universe that expands and later contracts under gravity. If the density of matter were high enough, gravitational attraction could eventually overcome expansion, leading to a cosmic collapse often called the Big Crunch.

In early cyclic proposals, the idea was simple. The universe expands from a Big Bang, slows down, reverses into a Big Crunch, and then rebounds into another Big Bang. Expansion and contraction repeat indefinitely.

This would mean that what we call the beginning might simply be the aftermath of a previous cosmic cycle.

The Entropy Problem

The greatest challenge to cyclic models comes from the very principle we examined earlier. The Second Law of Thermodynamics states that entropy increases in an isolated system. If the universe is isolated, its entropy should rise with each cycle.

Let us imagine a first cycle. The universe begins in low entropy, evolves, forms stars and galaxies, and then collapses. By the time collapse occurs, entropy is far higher than at the beginning.

Now, suppose a bounce occurs and a second expansion begins. That second cycle would not start in the pristine low entropy state of the first. It would inherit the entropy of the previous cycle. With each repetition, entropy would accumulate.

The result would be that each new cycle becomes larger and longer. Eventually, cycles would stretch so much that the universe would effectively approach heat death anyway. The rhythm would fade into stillness.

This entropy accumulation argument was historically one of the strongest objections to simple cyclic models.

In recent decades, more sophisticated cyclic models have been proposed. One of the most notable is the ekpyrotic model developed by Paul Steinhardt and Neil Turok.

In this framework, our observable universe may exist on a three-dimensional membrane embedded in a higher-dimensional space *(Brane Cosmology)*. Periodically, this membrane collides with another. Each collision releases immense energy, creating conditions similar to what we call the Big Bang.

Instead of a singular explosion from nothing, the Big Bang becomes a transitional event in a larger cosmic structure.

In such models, entropy can be diluted during periods of accelerated expansion driven by dark energy. If space stretches enormously before the next cycle, matter becomes so diluted that the next bounce can begin in a smoother, lower effective entropy state. The entropy is not destroyed, but it becomes so spread out that locally the new cycle appears ordered again.

Saying more on this topic is not preferable because whether this mechanism truly solves the entropy problem remains debated, and cyclic universe models are neither a fantasy nor a fact.

Denouement II

You might have expected a simple answer. A clear statement declaring whether time is real, whether it loops, whether it ends, whether the universe repeats like a heartbeat or fades like an echo. But time refuses to be reduced to a sentence.

What we have discovered instead is deeper.

Time is neither an illusion nor a rigid arrow fixed eternally in one direction. It is a consequence. A consequence of how matter arranges itself. A consequence of how energy spreads.

But this realisation leads us naturally to a new question.

If entropy governs time through the arrangement of matter and energy, then what exactly is this matter? What forms does it take? From colossal galaxy clusters stretching across millions of light-years to subatomic particles flickering into and out of existence, the universe expresses itself across an astonishing range of scales.

And so we move forward.

From the biggest to the smallest...

3

The Biggest To The Smallest

"How fascinating it is that something ten billion light-years wide is fundamentally composed of something only 10^{-10} meters small.

"

For decades, atoms were considered the smallest building blocks of matter. They were thought to be indivisible, fundamental, complete. Then science advanced. Experiments revealed that atoms themselves were composed of electrons, protons, and neutrons. Soon after, even protons and neutrons were found to contain smaller constituents called quarks. What once appeared final turned out to be another layer waiting to be opened.

A similar shift occurred when we looked upward instead of inward. For centuries, our Sun seemed unimaginably vast. It dominated our sky, powered our world, and defined our sense of cosmic scale. But when astronomers began studying distant galaxies through the powerful eye of the Hubble Space Telescope, they discovered stars far larger than our Sun. Among them was UY Scuti, a star so enormous that if placed at the centre of our Solar System, it would extend beyond the orbit of Jupiter. What once seemed supermassive became modest in comparison.

Each time we believed we had reached the limits of size, nature quietly revealed a greater extreme.

So a question remains.

How exactly does the scale range? From the minutest particles that flicker at the foundation of reality to the colossal structures that stretch across billions of light-years, what is the true spectrum of existence? How small can matter be, and how large can structure grow?

In this chapter, we will travel across that vast ladder of scale. We will descend into the microscopic depths and ascend into the cosmic heights.

Universe & Structures

You might ask what the answer is to how large the universe is. The honest answer is that we do not fully know. What we can measure is the observable universe. Because light travels at a finite speed, we can only see as far as light has had time to reach us since the beginning of cosmic expansion. That observable boundary forms a sphere around us.

Current measurements suggest that the observable universe has a diameter of about 93 billion light-years. This number may seem strange at first: if the universe is about 13.8 billion years old, how can we see objects 46 billion light-years away in each direction? The reason lies in expansion. Space itself has stretched while light was travelling. The galaxies we observe today are much farther away than their light travel time alone would suggest.

Beyond the observable universe may lie vastly more space. It could be finite yet unbounded, like the surface of a sphere in higher dimensions.

It could also be infinite. Observations cannot yet determine this conclusively.
But within this cosmic area we are bound, some structures challenge the imagination.

The universe is not randomly scattered matter. On enormous scales, galaxies gather into patterns that resemble a cosmic web. Filaments of galaxies stretch across hundreds of millions of light-years, separated by vast voids.

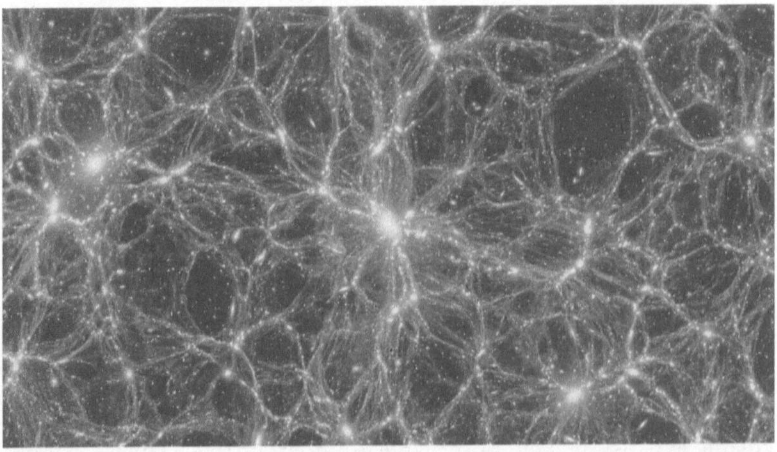

Figure 3.1: Visual Representation of Galaxy Clusters.

Among the largest known structures is the Hercules Corona Borealis Great Wall, a colossal grouping of galaxies spanning roughly 10 billion light-years. If confirmed as a single coherent structure, it challenges previous expectations about how large cosmic structures can grow without violating cosmological principles. There is also the Sloan Great Wall and the Huge Large Quasar Group, each extending across hundreds of millions to billions of light-years.

These are not solid objects. They are gravitationally linked patterns of galaxies. Yet in terms of sheer scale for us now, they are the largest known arrangements of matter.

Descending slightly in scale, we encounter galaxy clusters and superclusters.

Galaxy clusters are gravitationally bound collections of hundreds or thousands of galaxies. One of the most massive known clusters is El Gordo, formally known as ACT CL J0102-4915. It is a merging cluster containing enormous mass, including dark matter, hot gas, and galaxies.

Clusters themselves group into superclusters. Our own Milky Way belongs to the Laniakea Supercluster, which spans about 520 million light-years and contains around one hundred thousand galaxies.

If clusters are the cities of the universe, superclusters are the continents.

Among individual galaxies, size can be measured in terms of mass or diameter. One of the largest known galaxies is IC 1101. It is a supergiant elliptical galaxy located at the centre of a galaxy cluster. Its diameter is estimated to be about six million light-years, far larger than our Milky Way, which spans roughly one hundred thousand light-years.

IC 1101 contains trillions of stars. If our Sun were one grain of sand, the stars within such a galaxy would form the Saharan desert.

Galaxies themselves are not uniform disks. They are complex gravitational ecosystems, containing stars, gas, dust, black holes, and vast halos of dark matter. At their centres often lie supermassive black holes millions or billions of times more massive than the Sun.

Although black holes are compact, their mass can be enormous. The most massive known black holes are classified as ultramassive black holes. Some exceed tens of billions of solar masses; some black holes would engulf our entire Solar System if placed at its centre.

So let's talk about something which is actually placed at the centre of our solar system - the Sun.

For centuries, our Sun appeared enormous beyond comprehension. It dominates our sky, contains 99.8% of the mass of our Solar System, and powers life on Earth. Yet in the grand hierarchy of the cosmos, the Sun is only an average-sized star — almost modest when compared to the true giants of the universe. Before discussing the largest stars in the universe, it is important to understand that stars are not all the same. They differ in size, brightness, colour, temperature, lifespan, and even destiny. Some live quietly for trillions of years. Others burn fiercely and die young in violent explosions. Some collapse into dense remnants; others fade slowly into cosmic embers.

A star begins its life inside a vast cloud of gas and dust. Under the influence of gravity, this material slowly contracts. As it collapses inward, the core grows hotter and denser. When the temperature becomes high enough, hydrogen nuclei begin fusing into helium. This nuclear fusion releases tremendous energy. The outward pressure produced by this energy balances the inward pull of gravity, creating a stable, shining star. Astronomers call such stars 'main sequence stars'. So, if you wonder, which category does our sun fall in? Yes, it is a main-sequence star, and nearly ninety percent of all stars in the universe are currently in this stable stage. Main-sequence stars vary greatly in size and brightness. Some are much larger and hotter than the Sun, while others are significantly smaller and cooler.

Among the smaller ones are red dwarfs. These stars contain only a fraction of the Sun's mass and shine much more faintly. Because of their internal structure, they use their hydrogen fuel very efficiently, allowing them to burn slowly for extraordinarily long periods of time. Some red dwarfs may live for trillions of years, far longer than the current age of the universe. In fact, they are the most common type of star in our galaxy.

We talked about the dwarf, now the giant. When a star similar in mass to the Sun eventually exhausts the hydrogen in its core, the delicate balance between gravity and fusion changes. The core begins to contract, heating up further. At higher temperatures, helium starts fusing into heavier elements such as carbon. Meanwhile, the outer layers expand dramatically. The star grows much larger and cooler at its surface, taking on a reddish appearance. At this stage, it is known as a red giant. Theories suggest that billions of years from now, our Sun will expand into such a giant, possibly engulfing the inner planets.

As the red giant continues to evolve, it becomes unstable and sheds its outer layers into space. What remains is the dense core, now exposed. This remnant is called a white dwarf. Though roughly the size of Earth, it contains nearly the mass of the Sun, making it extremely dense. A white dwarf no longer produces energy through fusion; instead, it slowly cools over billions of years.

More massive stars follow a far more dramatic path. After exhausting lighter elements, they begin fusing heavier ones in successive stages. This process continues until iron forms in the core. Since iron fusion does not release energy, the star can no longer support itself against gravity. The core collapses suddenly, triggering a tremendous explosion known as a supernova. The collapsed core that remains may become a neutron star, an object so dense that a mass greater than the Sun's is compressed into a sphere only about twenty kilometres wide. In some cases, if the original star was even more massive, the collapse may form a black hole.

There also exist objects that fall between planets and stars. Brown dwarfs possess more mass than Jupiter but not enough to sustain long-term hydrogen fusion. They emit little visible light and occupy an intermediate position in the cosmic hierarchy.

Having discussed various kinds of stars, among all stars currently known to astronomy, the largest measured by radius is Stephenson 2-18. It is a red supergiant located nearly 19,000 light-years away from Earth in the constellation Scutum. It resides within a massive stellar grouping known as the Stephenson 2 cluster. What makes this star extraordinary is not merely its brightness or its mass, but its staggering size. Let us take a pause and visualise that. If Stephenson 2-18 were placed at the centre of our Solar System, its outer surface would stretch beyond the orbit of Saturn. Mercury, Venus, Earth, Mars, Jupiter — all would be engulfed within the star's fiery envelope. Even Saturn would orbit inside its glowing atmosphere.

Something that we left behind while discussing other things was the black holes. Something that we missed here is the black holes. Among all cosmic objects, black holes represent the most extreme consequence of gravity. They are not stars in the ordinary sense, nor are they merely dense remnants. They are regions of space where gravity becomes so strong that nothing, not even light, can escape once it crosses a certain boundary known as the event horizon. What we know is that a black hole forms from the death of a very massive star. But why? Because when such a star exhausts its nuclear fuel, fusion can no longer support it against its own gravity. The core collapses catastrophically. If the remaining mass is great enough, no known force can stop the collapse. Now, you might expect that what remains is a solid object, but actually, it is a region where spacetime itself is curved to an extreme degree.

Let us now know the types of black holes. Black holes are generally classified into several types based on their mass:

The first type is the stellar-mass black hole. These form from individual massive stars and typically range from a few to several dozen times the mass of the Sun. They are scattered throughout galaxies, often detected indirectly through their gravitational influence on nearby stars or gas.

The second type is the intermediate-mass black hole. These are heavier than stellar-mass black holes but lighter than the supermassive ones found in galactic centres. Their formation mechanism is still under investigation. They may form from the merging of smaller black holes or from the collapse of unusually large primordial stars in the early universe.

The third and most fascinating category is the supermassive black hole. These contain millions or even billions of times the mass of the Sun. Nearly every large galaxy studied so far appears to host one at its centre. Our own Milky Way contains a supermassive black hole known as Sagittarius A*, which has a mass of about four million Suns, and the decoded images describe its accretion disk as having a reddish appearance.

*Figure 3.2: Sagittarius A**

This leads to an important question: why are supermassive black holes found at the centres of galaxies? The answer lies in gravity and galactic evolution. During the formation of a galaxy, matter naturally flows toward the densest region due to gravitational attraction. The central region becomes a gravitational well where gas accumulates efficiently. If a black hole forms early in this region, it will continue to grow as more matter falls inward. Over time, a feedback relationship develops. The black hole influences the galaxy by regulating star formation through powerful jets and radiation, while the galaxy supplies material that feeds the black hole. In this way, galaxies and their central black holes evolve together.

To truly grasp this extremity, imagine taking Mount Everest, which has a mass of approximately 6.37×10^{14} kilograms. Though enormous by earthly standards, it is nowhere near naturally capable of forming a black hole. To turn Mount Everest into one, you would need to compress its entire mass into a radius of less than one attometer (about 10^{-15} meters) — roughly the size of a single atomic nucleus. Only at that incomprehensibly tiny scale would its escape velocity equal the speed of light. The result would not be something gigantic, but a microscopic black hole which, according to theoretical predictions involving Hawking radiation, would evaporate almost instantly. Why and How? The answer to this lies in the Schwarzschild Radius. It is the critical radius around any mass at which gravity becomes so strong that the escape velocity equals the speed of light. If a certain amount of mass or energy is compressed within this radius, the object becomes a black hole. The size of the radius depends on the amount of mass involved: the greater the mass, the larger the radius.

$r_s = 2GM / c^2$

Where:

r_s = Schwarzschild radius

G = Gravitational constant

M = Mass

c = Speed of light

However, on Earth, such compression is impossible, but in the universe, the violent deaths of massive stars naturally provide the conditions required for such extraordinary gravitational collapse. Now, what exactly is Hawking Radiation?

For a long time, black holes were thought to be completely black and eternal. However, in the 1970s, the physicist Stephen Hawking proposed something revolutionary. By applying quantum mechanics to the curved spacetime near a black hole, he showed that black holes are not entirely black. They can emit a faint radiation, now known as Hawking radiation.

There are many more intricacies of a black which we shall look upon later in Chapter '5' 'Empty or Not?'

Among the largest known black holes is TON 618. This extraordinary object is estimated to contain tens of billions of solar masses. It resides at the centre of a distant quasar, shining brightly as surrounding matter spirals inward and heats up before crossing the event horizon. Objects like this push the limits of how massive black holes can grow.

When we mentioned TON 618, we described it as residing at the centre of a quasar. To fully understand why TON 618 appears so extraordinarily bright, we must first understand what a quasar is.

A quasar is not a star. It is not even a single object in the traditional sense. A quasar is the extremely luminous core of a distant galaxy, powered by a supermassive black hole that is actively feeding on surrounding matter. They are among the most luminous, powerful, and energetic objects known in the universe. In fact, a single quasar can emit more light than hundreds of billions of stars combined. This is why TON 618 appears so brilliant.

Quasars were much more common in the early universe. Billions of years ago, galaxies contained abundant gas, allowing their central black holes to grow rapidly. Over time, as available fuel diminished, many quasars became quieter. Today, our own galaxy's central black hole, Sagittarius A*, is relatively inactive compared to those ancient luminous cores.

In continuity, we have nebulae, vast clouds of gas and dust where stars are born or remnants of stellar death. Some nebulae even stretch across hundreds of light-years. The Tarantula Nebula in the Large Magellanic Cloud is one of the most active star-forming regions known. It contains some of the most massive and luminous stars ever observed. If it were located as close to Earth as the Orion Nebula, it would cast visible shadows at night. Its core region, known as R136, hosts extremely massive young stars whose radiation and stellar winds shape the surrounding gas dramatically. While mentioning these cosmic structures, it is worth pausing to mention something spectacular: The Pillars of Creation.

Figure 3.3: Pillars of Creation
Source: NASA Science

These towering columns of gas and dust lie within the Eagle Nebula, a star-forming region about 7,000 light-years away. The pillars are several light-years tall and consist primarily of cold molecular hydrogen and dust. Within these dense columns, gravity is slowly pulling material together to form new stars. Intense radiation from nearby massive stars gradually erodes the pillars, sculpting them into their dramatic shapes.

Studying the most dense, massive, and spectacular structures of the universe draws us to a very curious thought that some of the largest features in the universe are not concentrations of matter but absences of it. Cosmic voids stretch across hundreds of millions of light-years, containing only a sparse scattering of galaxies where we might otherwise expect dense clusters and luminous structures. These voids form the negative space of what astronomers call the cosmic web.

On the grandest scales, the universe is not randomly distributed. Galaxies do not float uniformly in every direction. Instead, they are arranged in an intricate network of filaments and nodes. Long, thread-like structures of dark matter and galaxies stretch across space, intersecting at dense clusters. Between these filaments lie enormous regions where matter is strikingly scarce. These are the cosmic voids.

A cosmic void is not completely empty. It still contains dark matter, faint gas, and the occasional lonely galaxy. But compared to galaxy clusters, its density is extraordinarily low. Imagine the structure of foam or a sponge: solid walls enclosing vast pockets of emptiness. One well-known example is the Boötes Void, an immense region roughly 330 million light-years across. When first studied in detail, astronomers were surprised by how few galaxies it contained. Compared to the surrounding cosmic web, it appeared almost barren. Yet even such a "void" still hosts some galaxies, reminding us that emptiness in the universe is relative, not absolute.

The Measurement of The Huge

Studying these massive cosmic structures, a simple yet curious question might arise in your head: how do we measure something that is thousands or even billions of light-years away? How do we estimate the size of a star that we cannot visit, the mass of a black hole we cannot see directly, or the distance to a galaxy whose light began its journey before Earth even formed?

For a considerable period, UY Scuti was regarded as the largest known star. Its estimated radius made it the benchmark for stellar size in popular science and textbooks. However, with improvements in observational instruments and data analysis, more refined measurements suggested that Stephenson 2-18 may exceed it in radius. The shift did not occur because the universe changed, but because our tools did. As telescopes become more precise and stellar models more accurate, earlier estimates are revised. What appears definitive at one time often becomes provisional with better technology.

The answer to it lies in light - another most fascinating phenomenon in physics. Everything we know about distant cosmic objects comes from the electromagnetic radiation they emit or influence. Visible light is only a small portion of this spectrum. Astronomers also observe radio waves, infrared radiation, ultraviolet light, X-rays, and gamma rays. Each band reveals different physical processes.

Distances to nearby stars are determined using parallax, a geometric method based on Earth's motion around the Sun. For more distant stars and galaxies, astronomers use standard candles such as Cepheid variables and certain types of supernovae, whose intrinsic brightness is known. By comparing true brightness with observed brightness, distance can be calculated. Mass is often determined through gravitational effects—by studying orbital motion in binary systems, the Doppler shift in spectral lines, or the movement of stars around invisible central objects like Sagittarius A*. Stellar sizes are inferred using luminosity, temperature, and physical laws such as the Stefan–Boltzmann relation.

These techniques involve advanced spectroscopy, interferometry, and relativistic modelling, each requiring complex mathematics and careful calibration. A full exploration would demand an entire volume on observational astrophysics. For our purpose, it is enough to understand that the "biggest" or "most massive" objects are measured through the consistent application of physical laws to the light and motion we observe.

From Stars To Quarks

So, when we began this journey, we spoke of stars swollen beyond planetary orbits, of black holes anchoring galaxies, of structures spanning millions of light-years. Now, let us descend step by step into the opposite extreme. From the blazing surfaces of red supergiants, from quasars and cosmic voids, we move inward toward the fabric of matter itself.

The same universe that builds galaxies also builds atoms. The same physical laws that govern stellar collapse govern the stability of a hydrogen atom. The scale changes dramatically, but the logic of nature remains consistent. Let us now move from stars to quarks.

An atom is the smallest, basic building block of matter that constitutes elements, participating in chemical reactions without being created or destroyed. Seems familiar, right? Yet what is often overlooked is how extraordinary atoms truly are. An atom is mostly empty space. If the nucleus of an atom were the size of a small marble placed at the centre of a cricket stadium, the electrons would be orbiting somewhere near the outer seating, and everything in between would be emptiness. About 99.99% of an atom's volume contains no solid matter at all. The solidity you feel when you touch a table arises not from tightly packed substance, but from electromagnetic forces preventing atoms from passing through one another. Atoms consist of a central nucleus surrounded by electrons. The nucleus contains protons and neutrons, while electrons exist in quantised energy levels around it. These levels are not tiny planetary orbits as once imagined; they are probability clouds described by quantum mechanics. An electron does not travel in a neat circular path—it exists as a wavefunction, a mathematical description of where it is likely to be found.

Another lesser-known fact is that atoms are ancient. Many of the atoms in your body were forged inside stars billions of years ago. Rooted in the principles of astrophysics and stellar nucleosynthesis, this concept reveals how the Big Bang gave rise primarily to the lightest elements—hydrogen and helium—while the heavier ones vital to life, including carbon, oxygen, nitrogen, and iron, were born in the searing cores of massive stars or amid the cataclysmic blasts of supernovae billions of years in the past. You are, quite literally, made of recycled stardust, a term popularised by the great astronomer Carl Sagan.

When atoms bond together, they form molecules. Molecules arise because electrons can be shared or transferred between atoms, creating stable configurations. The water molecule, for example, forms because oxygen shares electrons with two hydrogen atoms. The angle between those bonds gives water its unique properties. Carbon offers another astonishing example. The same element can form diamond, one of the hardest known materials, and graphite, which is soft enough to leave marks on paper. The difference lies only in how the carbon atoms are arranged. Structure determines property. Even more fascinating is that molecules vibrate constantly. The microscopic world is not static. It is dynamic, restless, and governed by probabilistic rules rather than rigid certainty.

Going deeper, inside every atom lies the nucleus, composed of protons and neutrons. Protons carry a positive electric charge. Neutrons carry no charge. Electrons carry a negative charge. The balance between protons and electrons determines an atom's electrical neutrality. The mass of the atom resides almost entirely in the nucleus.

Protons and neutrons are about 2,000 times heavier than electrons. If the electron were the mass of a mosquito, a proton would weigh as much as a small bird. Yet electrons play a decisive role in chemistry. They determine how atoms bond, how molecules form, and how matter interacts. One remarkable fact is that the mass of a nucleus is slightly less than the sum of the masses of its protons and neutrons. The missing mass has been converted into binding energy, according to Einstein's equation $E= mc^2$. This binding energy is what holds the nucleus together. It is also what powers nuclear reactions, including those inside stars.

Now, for decades, protons and neutrons were thought to be fundamental. But deeper experiments revealed that they are themselves composed of smaller particles called quarks.

A proton consists of two "up" quarks and one "down" quark.

A neutron consists of two "down" quarks and one "up" quark.

Quarks possess fractional electric charges—something never observed in isolation. An up quark carries +2/3 of the elementary charge, while a down quark carries −1/3. Combined, they produce the familiar charges of protons and neutrons.

Electrons, however, are different. They are considered elementary particles, meaning they show no internal structure in current experiments. They belong to a family called leptons.

As we descend further, we reach a boundary beyond which current physics struggles to describe reality: the Planck length.

Denoted as l_P, it is a fundamental unit of length in physics, representing the scale at which quantum effects of gravity become significant and our current theories—general relativity and quantum mechanics—break down, necessitating a unified theory of quantum gravity. It emerges from dimensional analysis as the unique combination of three universal constants that yields a length: the reduced Planck's constant (ħ), Newton's gravitational constant (G), and the speed of light (c).

The exact formula is:

$l_P = \sqrt{(\hbar G / c^3)}$

This was first proposed by Max Planck over a century ago as part of the system of Planck units, which normalise fundamental constants to unity for theoretical convenience. Numerically, its value is approximately

1.616×10^{-35} meters. You can't even imagine this size; this is unimaginably small. Far smaller than a proton, far smaller than any scale we can directly probe. Some theories suggest that at the Planck scale, spacetime becomes "foamy," fluctuating due to quantum uncertainty. The Planck length is not merely small, but it represents a limit. Below this scale, our current physical theories lose predictive power.

Let us discuss something we did not directly after the quarks. As we descend beyond quarks and leptons, beyond the familiar framework of the Standard Model, we encounter one of the most ambitious ideas ever proposed in theoretical physics: string theory. According to this idea, the fundamental constituents of reality are not point-like particles but unimaginably tiny one-dimensional strings, vibrating at specific frequencies. These strings would be so small that their length is thought to be on the order of the Planck scale, far beyond any current experimental reach. What we perceive as different particles—electrons, quarks, neutrinos—would not be separate building blocks in the traditional sense, but different vibrational modes of the same underlying entity, much like different musical notes arise from different vibrations of a single violin string. In this framework, mass, charge, and other particle properties emerge from how these strings vibrate. One of the most remarkable features of string theory is that it naturally includes a quantum description of gravity; a particular vibrational mode corresponds to the hypothetical graviton, the particle that would mediate gravitational interaction, something the Standard

Model does not successfully incorporate. Furthermore, consistent versions of string theory require additional spatial dimensions beyond the familiar three, suggesting that the universe may contain hidden dimensions compactified at extremely small scales. Which we had discussed in our first chapter. While there is currently no direct experimental evidence confirming the existence of strings, the theory remains compelling because of its mathematical consistency and its potential to unify all fundamental forces within a single framework. If true, it would mean that at the deepest level, reality is not built from solid particles but from tiny vibrating filaments of energy, and that the diversity of matter arises from variations in vibration rather than variations in substance.

The Measurement of The Minute

Now, whether a deep question about how these extremely small particles are measured arises in your head or not. I'll tell you the answer!

Atoms were first inferred through chemical behaviour and later visualised indirectly using scanning tunnelling microscopes. Electron properties are measured through electromagnetic interactions and spectroscopic analysis. To probe inside protons, scientists use particle accelerators. Facilities such as the CERN (The European Organisation for Nuclear Research) accelerate particles to near light speed and collide them. By analysing the debris of these collisions, physicists infer the presence of quarks and other subatomic particles. The Large Hadron Collider is currently the most powerful such machine. It allows scientists to explore energy scales where new particles can emerge. Measurements at these scales rely on detectors that track particle trajectories, energy deposition, and decay patterns. Nothing is seen directly; everything is inferred from interaction signatures. Even the Planck length is not measured directly. It arises from theoretical reasoning based on fundamental constants.

Denouement III

So, we have travelled from stars spanning billions of kilometres to quarks confined within atomic nuclei. We have moved from cosmic voids to Planck-scale speculation. In doing so, we have seen that the universe is continuous across scales. The same physical principles echo from the largest structures to the smallest.

Yet what we have discussed here is only the surface.

We have mentioned quarks, leptons, gluons, and quantum fields, but we have not explored them in depth. We have spoken of fundamental forces without fully understanding their unification. We have touched upon the Planck scale without unravelling the mysteries it holds.

In the next chapter, we will go deeper into the world of fundamental particles. We will examine the Standard Model, explore the nature of forces, and question whether quarks and leptons are truly the final layer—or whether something even more fundamental awaits discovery.

From stars to quarks, the journey narrows.

From quarks onward, it becomes even more profound...

4

Cat In A Box

"Infinite spots lie hidden, yet only one is seen!"

If I ask you whether an electron is a particle or a wave, what would your answer be? If we rely on the conventional physics we have studied, the physics of Newton, of moving tennis balls and planets, you would confidently say it is a particle. After all, electrons have mass, they carry a charge, they can collide, scatter, and be counted. Everything about our everyday reasoning pushes us toward that conclusion. But nature is not obligated to obey our intuitions.

When scientists began performing experiments at the atomic scale, especially the famous double-slit experiment, the story changed dramatically. In that experiment, electrons were fired toward a barrier containing two narrow openings. If electrons were simple particles, we would expect them to pass through one slit or the other, producing two distinct clusters on a screen behind the barrier. Instead, what appeared was something very astonishing. To explain this strange behaviour, physicists developed the framework of quantum mechanics.

At its heart lies the principle of superposition, which is the idea that a quantum system can exist in multiple possible states simultaneously until it is measured. If microscopic objects can exist in multiple states at once, what prevents the same from happening to larger objects? Where exactly does the boundary between the quantum world and the classical world lie? Does observation create reality? Does the act of measurement collapse possibilities into a single outcome? Or are all possibilities realised in different branches of existence? All these questions we will answer later in the chapter, but before answering them, let us take a look at something fascinating.

To highlight this paradox, the physicist Erwin Schrödinger proposed a thought experiment called "Cat In A Box."

Alive Or Dead?

The question sounds simple. The cat inside the box is either alive or dead. In everyday life, this question would never disturb us. A cat is breathing, or it is not. A coin tossed in the air lands heads or tails. A switch is on or off. Classical physics supports this intuition. Objects possess definite properties whether or not we observe them. The universe, in that familiar picture, is concrete and predictable. Everything seems to be quite simple until Quantum mechanics challenges that certainty.

To understand why the cat in the box becomes a scientific and philosophical puzzle, we must first understand what quantum mechanics says about microscopic systems. In classical physics, a particle has a definite position and velocity at every moment. Even if we do not know those values precisely, we assume they exist. If you see the blackboard in your class and close your eyes, it doesn't vanish; it stays right there. Quantum mechanics replaces that certainty with probability.

In quantum mechanics, certain physical systems do not possess definite values for particular properties before measurement. Instead, they are described by a wavefunction, a mathematical object that encodes all possible outcomes and the probability of each. The system exists in a superposition of states. This means that multiple possibilities coexist simultaneously in the mathematical description.

Let me tell you that superposition is not a poetic metaphor. It is a precise feature of the equations governing quantum systems. If a radioactive atom has a fifty percent chance of decaying within an hour, quantum theory does not describe it as secretly decayed or secretly intact while we are unaware. Rather, before measurement, it is described as a combination of both possibilities. This is where the thought experiment proposed by Erwin Schrödinger becomes powerful.

Imagine a sealed box containing a cat, a tiny amount of radioactive material, a detector that senses decay, and a vial of poison connected to the detector. If the atom decays, the detector triggers the release of poison, and the cat dies. If the atom does not decay, nothing happens, and the cat remains alive. Radioactive decay is governed by quantum probability. Before observation, the atom is described as being in a superposition of decayed and not decayed states. If we apply quantum mechanics consistently, the detector becomes entangled with the atom. The poison mechanism becomes entangled as well. Finally, the cat becomes entangled with the entire chain of events. The combined system evolves into a state that includes both the atom decayed with the cat dead and the atom not decayed with the cat alive.

Figure 4.1: A Cat in a Box With a Radioactive Substance, Vial of Poison and Radiation Detector
Source: Sciencenotes.org

In strict quantum language, the cat is in a superposition of alive and dead. This conclusion feels absurd. We never encounter animals that are partly alive and partly dead. When we open the box, we see a definite outcome. The cat is one or the other.

So what happens between the mathematical description and our observation? One way to approach this is through analogy. Imagine tossing a coin in a completely dark room. You hear it land, but you cannot see the result. Until you switch on the light, you do not know whether it shows heads or tails. In classical thinking, the coin has already landed in one definite state. Your ignorance does not affect reality. It only limits your knowledge.

Quantum mechanics differs fundamentally. In the quantum analogy, the coin is not simply hidden in darkness. It exists in a state that genuinely includes both heads and tails until observed. Observation does not merely reveal the result. It plays a role in selecting it.

Another analogy involves a musical chord. When multiple notes are played together, they form a combined sound. In quantum mechanics, superposition resembles such a combination. The system contains several possible states simultaneously. Measurement isolates one component from that combination.

However, analogies can mislead. In the case of the coin, classical uncertainty is about ignorance. In quantum mechanics, uncertainty is built into the nature of reality itself. The mathematical framework of quantum mechanics is precise. The wavefunction evolves smoothly and deterministically according to the Schrödinger equation. Yet measurement introduces probability. The probability of each outcome is given by the square of the amplitude of the wavefunction component corresponding to that outcome:

$P = |\langle \text{outcome} \mid \psi \rangle|^2$

This is known as the Born Rule. Here, ψ is the wave function (or quantum state) of the system before measurement, $\langle \text{outcome} \mid \psi \rangle$ is the complex probability amplitude which is a number between -1 and 1 in magnitude, and$|\ldots|^2$ means you take the absolute value (magnitude) of that complex number and square it to get the real probability which will always between 0 and 1. This squaring is crucial. It allows interference to occur because amplitudes can add constructively or destructively before being squared. This formula explains why quantum mechanics is not vague speculation. The theory does not say "anything can happen." It assigns precise numerical probabilities to each possible result. If the wavefunction contains equal contributions from two states, then each outcome has equal probability. If one contribution is larger, its corresponding outcome is more likely. The mathematics predicts the statistical distribution of results with extraordinary accuracy, as confirmed by countless experiments.

Let us frame it in the context of the cat in the box, the wavefunction evolves into a superposition containing both the atom decayed with the cat dead and the atom not decayed with the cat alive. The Born rule tells us the probability that, upon opening the box, we will observe one state rather than the other. If the decay probability in one hour is fifty percent, then the squared amplitudes of the two components are equal, and the chances of finding the cat alive or dead are each one-half.

The important point is this: before measurement, the wavefunction contains both possibilities. After measurement, one definite outcome is observed, with probability determined by the square of the amplitude. The formula does not describe a gradual blending into reality. It describes a probabilistic selection governed by a precise mathematical structure.

The deep issue here is the measurement problem. According to the Schrödinger equation *(not the Born Rule)*, the wavefunction evolves smoothly and deterministically over time. There is no sudden jump built into this equation. Yet when measurement occurs, we observe a single definite result. The wavefunction appears to collapse from a combination of possibilities into one outcome. So, what causes this collapse? One traditional explanation is known as the Copenhagen interpretation. In this view, the wavefunction represents our knowledge of the system. Measurement forces the system into a definite state, and the collapse is a real physical process. The classical measuring device plays a special role in defining outcomes.

However, this raises further questions. Where exactly does the boundary between the quantum system and the classical apparatus lie? The detector inside the box is made of atoms. The poison mechanism is made of atoms. The cat is made of atoms. If quantum laws apply universally, there should be no sharp dividing line.

Another approach involves decoherence. When a quantum system interacts with its environment, it becomes entangled with countless surrounding particles. These interactions rapidly destroy the coherent relationships between different components of the superposition. As a result, interference effects vanish, and the system behaves as though it has chosen a definite state. Decoherence explains why macroscopic superpositions are not observed in daily life. The environment constantly monitors large systems, preventing quantum ambiguity from persisting. Yet decoherence does not entirely eliminate the conceptual puzzle. It explains why we do not observe interference between alive and dead states, but it does not specify why only one outcome is experienced.

A more radical and fascinating interpretation is the many-worlds interpretation. In this framework, the wavefunction never collapses. Every possible outcome actually occurs. When the box is opened, the universe branches. In one branch, the observer sees a living cat. In another branch, the observer sees a dead cat. Both outcomes exist, but in separate, non-interacting branches of reality.

This idea may seem extraordinary, but it follows logically from taking the Schrödinger equation seriously without adding collapse. The universe continually splits into multiple histories, each corresponding to a different quantum possibility.

Another important concept tied to the cat experiment is entanglement. When two systems interact, their states can become inseparable. The state of one cannot be described independently of the other. In the box, the atom, detector, poison, and cat form a single entangled system. The fate of each component is linked to the others. Entanglement has been demonstrated experimentally and forms the foundation of emerging technologies such as quantum communication and quantum computation. It is not merely philosophical speculation. It is a measurable feature of nature.

Getting the 'Cat In A Box' theory very well, we can now look at something more interesting, which will fetch us the answer to our question.

Particle Or Wave

And now we step into one of the most astonishing and carefully tested experiments in the history of science: the electron double slit experiment. This experiment is not merely a laboratory curiosity. It is a turning point in our understanding of reality. It challenges the classical picture of matter and opens the doorway to quantum mechanics in its full depth.

Let us imagine a barrier with two narrow slits cut into it. Behind the barrier is a detection screen that can record where incoming objects land. In front of the barrier, we have a source that fires tiny particles toward the slits. In classical physics, if we fire small solid objects such as tiny beads or grains of sand at a barrier with two openings, what do we expect? Each particle travels along a definite path. If a bead passes through the left slit, it lands on a particular region of the screen aligned with that slit. If it passes through the right slit, it lands in a region aligned with that slit. After firing many beads, we see two clusters forming on the screen. Two slits produce two bands. This is intuitive and predictable.

Now let us replace the beads with waves, such as water waves in a ripple tank or light waves shining through two narrow openings. When waves pass through two slits, each slit acts as a new source of waves. The waves spread out and overlap. Where the crests of two waves meet, they add together. Where a crest meets a trough, they cancel. This overlapping produces a pattern of alternating bright and dark bands on the screen. This pattern is called an interference pattern. It is the signature of wave behaviour. Interference arises because waves can superpose. Superposition means that when two waves occupy the same region of space, the resulting disturbance is the sum of their individual effects. Constructive interference occurs when waves reinforce each other, creating bright regions. Destructive interference occurs when they cancel each other, creating dark regions. So far, nothing is shocking. Particles make two bands. Waves make interference patterns.

Now comes the experiment that unsettles everything. Instead of beads or light, we use electrons. According to classical physics, electrons are particles. They have mass. They carry an electric charge. They can be deflected by electric and magnetic fields. They leave tiny dots on detection screens. Everything about them suggests that they are small, localised objects. If electrons are particles, then firing them at two slits should produce two bands on the screen, just like beads. But when scientists experimented carefully, they did not observe two bands. They observed an interference pattern. Bright and dark fringes appeared on the screen. The electrons behaved as if they were waves passing through both slits and interfering with each other. This result alone is astonishing. But it becomes even more astonishing when we examine the experiment in detail. Perhaps the electrons were interacting with each other, creating wave-like behaviour collectively?

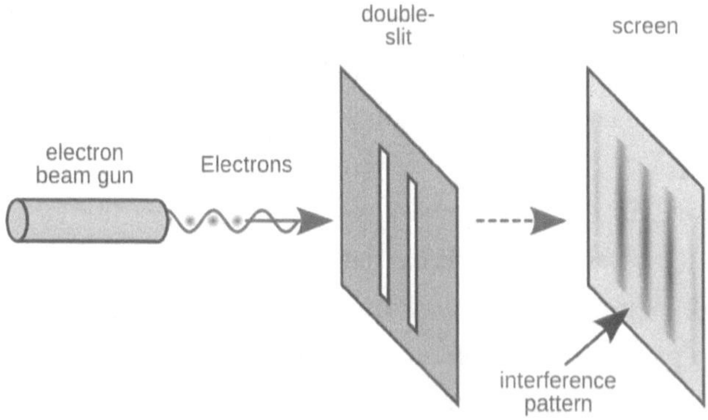

Figure 4.2: Electron Double-Slit Experiment.
Source: Wikipedia.org

To test this, physicists reduced the intensity of the electron beam so that electrons were fired one at a time. Not two at once. Not many at once. One single electron, travelling toward the barrier. If electrons are particles, each one should pass through one slit or the other and land on the screen, gradually forming two bands. Instead, something remarkable happened. Each individual electron arrived as a tiny localised dot on the screen. It looked like a particle upon detection. But after many electrons had been sent one by one, the overall pattern that emerged was still an interference pattern. Each electron, taken alone, seemed to contribute to a pattern that requires wave behaviour. It is as if each electron travels through both slits simultaneously and interferes with itself.

This leads us to the concept of the wave function. In quantum mechanics, a particle is described not simply by position and velocity, but by a wave function, usually denoted by the symbol psi. The wave function encodes the probability amplitudes for finding the particle in various locations, as discussed earlier. The wave function spreads out in space like a wave. When it encounters two slits, it splits into two parts. These parts propagate through both slits and overlap on the other side. Their superposition produces regions of high and low probability.

Therefore, in classical physics, yes, you may call the electron a particle. It possesses mass, it carries charge, and it can collide with other objects much like a speck of matter. Within that framework, calling it a particle is not only acceptable but practical. But in quantum mechanics, it is neither simply a particle nor merely a wave. It is fundamentally a wave function whose measurable manifestations can appear particle-like or wave-like depending on how we observe it.

Now, let us take a previously introduced analogy. Imagine you are listening to two musical notes played simultaneously. If they are in harmony, the sound becomes stronger. If they are out of phase, they can partially cancel. Similarly, probability amplitudes behave like waves. They can reinforce or cancel before being translated into measurable probabilities.

However, a deeper question can be asked. What if we try to observe which slit the electron passes through? Physicists placed detectors near the slits to determine whether the electron goes through the left or the right slit. When such measurements are made, the interference pattern disappears. Instead, two bands appear on the screen.

By attempting to gain information about the path, we destroy the interference. This phenomenon is related to the principle of complementarity introduced by Niels Bohr. According to this principle, quantum systems exhibit either particle-like behaviour or wave-like behaviour depending on the experimental setup. The two aspects are complementary. Observing one aspect excludes the other. The act of measurement plays a fundamental role. It is not merely a passive observation. It changes the system. This leads us to the concept of coherence. For interference to occur, the two parts of the wave function must maintain a definite phase relationship. When we measure which slit the electron passes through, we disturb this phase relationship. The wave function becomes entangled with the measuring device, and the coherence between the two paths is lost. This process is known as decoherence, which we had discussed earlier in short.

Decoherence also explains why interference patterns disappear when information about the path is available. The system becomes correlated with the environment, and the superposition effectively breaks down into classical alternatives. Now, you might or might not get the difference between decoherence and the Copenhagen interpretation. If you got it, fine, and if not, that's fair because they share a vivid similarity. Both the Copenhagen interpretation and decoherence agree that quantum superpositions become effectively unobservable in the macroscopic world, resulting in the appearance of definite classical outcomes. The key difference between the Copenhagen interpretation and decoherence lies in how they account for the transition from quantum superposition to the definite, classical-looking outcomes we observe.

The Copenhagen interpretation posits a fundamental, irreducible process called wave-function collapse: when a measurement occurs, the wave function instantaneously jumps from a superposition of possibilities to just one definite state. This collapse is not described by the Schrödinger equation—it is an extra postulate, triggered by the act of observation or interaction with a classical measuring device. While decoherence, by contrast, does not introduce any collapse or special measurement postulate. It explains the appearance of definite outcomes as a natural consequence of unitary evolution (Schrödinger equation) when the quantum system becomes entangled with its large, complex environment. Through this entanglement, the phase relationships required for observable interference are rapidly scrambled and effectively irreversible for all practical purposes. The system then behaves as if it were in a classical mixture of definite states, even though the full quantum state of system + environment remains in superposition. Decoherence thus provides a dynamical, physical mechanism for why superpositions become unobservable in the macroscopic world, without needing to invoke collapse or a privileged observer.

Another helpful analogy is to imagine a person walking through a maze with two corridors. In classical thinking, the person chooses one corridor. In quantum thinking, the wave function explores both corridors simultaneously, and the resulting interference determines the probability of emerging at different exits. Quantum computing, a very vast field, can be lightly explored by us as it harnesses the very principles illustrated by our maze analogy to solve certain problems far more efficiently than classical computers. In classical computing, a processor must explore paths through a computational "maze" (such as searching a database, factoring large numbers, or optimising complex systems) one route at a time, trying each possibility sequentially or in parallel, but still limited by the number of processors. A quantum computer, however, uses qubits that exploit superposition to explore many—often exponentially many—corridors simultaneously, as if the wave function of the entire computation spreads through all viable paths at once. Through carefully designed quantum gates and operations, these paths interfere with one another: amplitudes for incorrect or suboptimal routes cancel out (destructive interference), while those leading to the correct or optimal solution reinforce each other (constructive interference). When the computation concludes, and the qubits are measured, the probability is dramatically skewed toward the desired outcome, much like how the interference pattern in the double-slit experiment concentrates detections at certain spots.

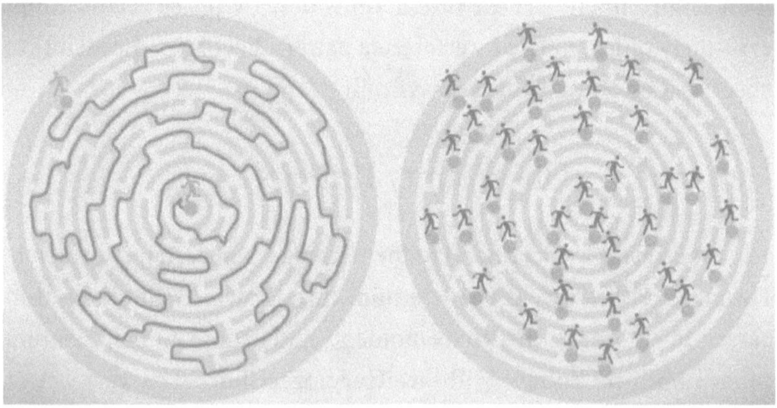

Figure 4.3: Visual Representation of Maze Solving by Bits vs. Qubits. Source: link.springer.com Artificial Intelligence in Medicine

This interference-driven amplification, rather than brute-force checking of every path, is what enables quantum algorithms (such as Shor's for factoring or Grover's for search) to achieve exponential or quadratic speedups on specific problems, turning what would take classical computers billions of years into minutes or hours—though only for tasks where the maze's structure allows such clever wave-like orchestration to shine.

The electron double slit experiment has been repeated countless times with increasing sophistication. It has been performed not only with electrons but also with neutrons, atoms, and even large molecules composed of many atoms. Each time, interference patterns emerge as long as coherence is preserved. This tells us something profound. The wave nature is not limited to tiny particles. It is a universal feature of quantum systems. Why do we not see such interference in everyday objects like footballs or cars? I think you might have guessed that the answer lies in decoherence. Large objects interact continuously with their environment. These interactions destroy phase relationships extremely quickly. The interference pattern becomes impossible to observe. The object behaves classically.

Heisenberg's Uncertainty Principle

The experiment also connects to the Heisenberg uncertainty principle. The more precisely we try to determine through which slit the electron passes, the more we disturb its momentum. Position and momentum cannot both be known with arbitrary precision. The very act of measuring position introduces uncertainty in momentum, affecting the interference pattern. And if you wonder what Heisenberg's Principle of Uncertainty says? Let me tell you: When we first encounter uncertainty in daily life, it usually means ignorance. If you do not know where your keys are, the uncertainty exists only in your mind. In classical physics, uncertainty arises from incomplete knowledge. If we had perfect instruments and perfect information, we could, in principle, measure both the exact position and exact momentum of any object at the same time.

Quantum mechanics overturns this assumption. In the microscopic world, uncertainty is not a limitation of our tools. It is a fundamental feature of nature itself. The Heisenberg Uncertainty Principle, formulated by Werner Heisenberg in 1927, states that certain pairs of physical quantities cannot be simultaneously known with unlimited precision. The most famous pair is position and momentum. Momentum, in simple terms, is mass multiplied by velocity. Position tells us where something is. Classical intuition tells us there should be no problem in measuring both precisely. After all, for a moving car, we can measure its location on the road and its speed at the same time.

But at the quantum scale, reality behaves differently.

The mathematical statement of the uncertainty principle is:

$\Delta x \cdot \Delta p \geq \hbar / 2$

Here, Δx represents the uncertainty in position, Δp represents the uncertainty in momentum, and $\hbar$ is the reduced Planck constant, which equals Planck's constant divided by 2π. The symbol $\hbar$ sets the scale at which quantum effects become significant.

This inequality does not mean that position and momentum cannot both be measured. It means that the product of their uncertainties can never be smaller than a certain minimum value. If you reduce the uncertainty in position, the uncertainty in momentum must increase. If you try to reduce the uncertainty in momentum, the uncertainty in position must increase.

This is not a technological limitation. Even with a perfect measuring device, this limit cannot be surpassed. It is woven into the structure of quantum theory.

To understand why this happens, we must recall that quantum objects are described by wave functions. A wave function spreads out in space and contains information about probabilities. A sharply localised position corresponds to a wave that is tightly confined. However, a tightly confined wave is composed of many different wavelengths superposed together. And different wavelengths correspond to different momenta.

This is a direct consequence of wave mathematics. A perfectly sharp spike in space would require an infinite range of wavelengths. Since wavelength is related to momentum, this means an infinite uncertainty in momentum. Conversely, a single, pure wavelength corresponds to a definite momentum, but such a wave extends infinitely in space, meaning the position is completely uncertain. The uncertainty principle, therefore, emerges naturally from the wave nature of quantum systems. It is not an added rule but a mathematical necessity.

Now, let me make you interestingly curious about something. We, in short, know that decoherence occurs when an observer detects the wave function. But how does that particle or wave function get to know about that observer? If you want to get an answer, we can connect the uncertainty theory with the experiment, and very easily, we will get our answer.

Let us take a very simple analogy first, and then we will look at the main target. A very simple and direct analogy for the Heisenberg Uncertainty Principle is the idea of trying to locate a small insect in a completely dark room using a torch.

Imagine a tiny insect flying inside a pitch-dark room. You want to know exactly where it is and how fast it is moving. Since the room is dark, the only way to see it is to shine light on it. If you use a very soft, dim light, the insect will barely be disturbed. However, the light is too faint to clearly determine its exact position. You only get a vague idea of where it might be. In this case, the disturbance is small, but the position is uncertain. Now, suppose you use a very powerful, focused flashlight to pinpoint its exact location. The strong beam strikes the insect and pushes it slightly. Because light carries momentum, the impact changes the insect's motion. You now know precisely where it was at that moment, but its speed and direction have been altered. The more precisely you try to fix its position, the more you disturb its motion.

Similarly, suppose we try to determine through which slit an electron passes. To do this, we must interact with it. We might attempt to illuminate it with light or use some kind of detector at the slit. But any interaction capable of revealing position must involve an exchange of energy or momentum.

If we shine light on the electron, the photons of light carry momentum. When a photon scatters off the electron, it transfers some of that momentum. Even if the photon is extremely weak, the momentum transfer cannot be made arbitrarily small if we wish to localise the electron precisely.

The more precisely we determine which slit the electron goes through, the more sharply we confine its position. And as the uncertainty principle tells us, the sharper the position, the larger the uncertainty in momentum. This disturbance in momentum alters the electron's subsequent motion and destroys the delicate interference pattern on the screen. In this way, the disappearance of interference is not merely due to clumsy measurement. It is a direct reflection of quantum law. The interference pattern requires a well-defined phase relationship between probability amplitudes associated with both slits. Once we localise the electron at one slit, the act of localising the electron increases Δx precision, which necessarily increases Δp uncertainty. This increased uncertainty alters the conditions required for interference. As a result, the interference pattern fades or disappears.

Denouement IV

Throughout this chapter, we have wrestled with one of the most unsettling truths of modern physics: certainty dissolves at the quantum scale. We began with a simple yet disturbing image, a cat sealed inside a box, suspended between life and death until observation forces a resolution. We journeyed through the strange logic of superposition, explored probability amplitudes, and encountered the mathematical precision of the Born Rule, which reminds us that what we measure emerges from the structure of the wave function itself. We examined the electron double slit experiment and saw interference patterns arise and vanish depending on whether we choose to observe the path. We confronted the Heisenberg Uncertainty Principle and learned that position and momentum cannot both be pinned down with unlimited precision. The deeper we try to grasp one, the more the other slips away. Moreover, we got the answers to the questions asked at the start of the chapter.

But now let us pause.

Imagine a sealed box from which every particle has been removed. Absolute vacuum. Classical reasoning would suggest complete silence, complete inactivity. Yet quantum principles whisper another possibility. If energy and time are connected through uncertainty, can energy within a region remain exactly zero at every instant? Or must there be subtle fluctuations, brief and fleeting, too small and too rapid to grasp directly?

We began this chapter questioning whether a microscopic system could exist in a superposition of mutually exclusive states. Now we prepare to ask a new question. When we say a space is empty, what do we truly mean? Is emptiness merely the lack of matter, or is it a dynamic state whose quietness hides activity too subtle for casual inspection?

5

Empty Or Not?

"Nothing is really empty; there can be something made out of nothing."

Imagine holding a glass in your hand. You turn it upside down. You shake it gently. Nothing falls out. You look inside, and it appears empty. Now imagine a much larger glass, not made of transparent material, but formed by the boundaries of space itself. Remove the air. Remove the dust. Remove every visible speck. Remove even the tiniest trace of gas. Seal it perfectly.

Now, what remains inside? Your first instinct may be to answer quickly. Nothing remains. That is what empty means: no matter, no particles, no objects. Just void.

But pause for a moment. Throughout our journey so far, we have learned that reality does not surrender its secrets to first impressions. The cat in the box was not simply alive or dead. The particle passing through the slits did not travel along a single neat path. Position and momentum refused to submit to perfect measurement. The certainty, when examined closely, fractured into probabilities.

So, in this curious journey of ours, let us explore another simple-looking yet fascinating question. Is a vacuum truly empty? We will answer this question later in the chapter. For now, let us allow the question itself to breathe. In classical thinking, space was often imagined as a passive stage. It was the silent background upon which matter moved. Planets revolved, apples fell from the trees, and light travelled almost instantaneously. But space itself did nothing. It simply existed as an arena, neutral and inactive. If all matter were removed from a region, classical reasoning would suggest that what remains is pure emptiness; you may call it a blank canvas or a void without structure. Yet modern physics has repeatedly challenged our most comfortable assumptions. With each discovery, something once taken for granted was revealed to be incomplete. And now we are confronting emptiness to understand what actually true emptiness means.

Let me tell you this, laboratories have achieved vacuums so pure that only a handful of atoms remain in volumes larger than entire rooms. But is that the end of the story? Obviously, No. Even if we imagine a region stripped of all matter, we are still left with space. And space, according to modern theory, is not merely a geometric abstraction. It is intertwined with time. It bends. It stretches. It carries properties that influence motion and energy. If space can curve in the presence of mass, as described by relativity, then it cannot be a mere absence. It must possess structure. And if quantum mechanics governs all physical systems, then it must also govern space itself. This realisation unsettles the simple image of emptiness as nothingness.

And not only this, but there is a quiet boldness in asking this question also. It challenges a concept so basic that we rarely examine it. We speak casually of empty rooms, empty boxes, empty containers. We imagine a vacuum as the absence of everything. But in physics, definitions matter.

A vacuum is typically defined as a region with no particles. Yet particles themselves, as we have learned, are not simple little marbles. They are manifestations of deeper underlying descriptions. They arise from fields. They are excitations of something more fundamental.

If particles can emerge from deeper structures, what remains of those structures when no particles are present? Do not worry, we will explore the nature of the vacuum carefully and patiently. We will examine what modern theory suggests about regions devoid of matter. We will investigate whether emptiness can remain perfectly calm or whether subtle processes continue unseen. But before investigating what something devoid of matter is, let us examine the fabric of matter carefully first.

The Fabric

This segment is called The Fabric because what we are about to examine is not merely a list of particles. It is the structural design of reality—the underlying texture from which everything tangible emerges, a tapestry woven from mathematical precision and cosmic necessity. This segment is called The Fabric because what we are about to examine is not merely a list of particles. It is the structural design of reality—the underlying texture from which everything tangible emerges, a tapestry woven from mathematical precision and cosmic necessity.

The Standard Model of particle physics, our most refined map of the microscopic world, presents a strikingly organised picture of nature. According to it, the universe is built from a specific set of fundamental ingredients. It predicts seventeen fundamental particles:

Twelve fermions, which form the building blocks of matter.

Five bosons, which mediate the fundamental forces.

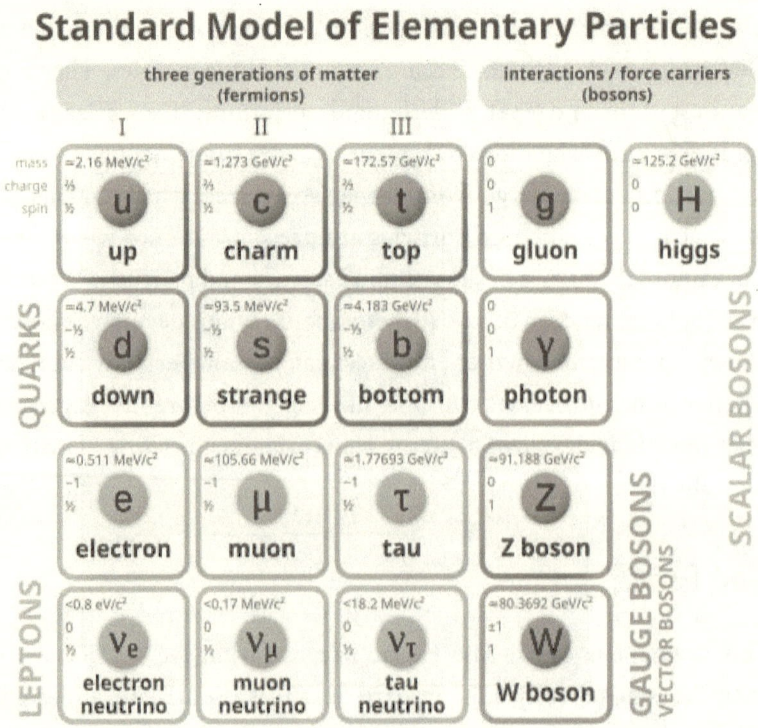

Figure 5.1: Standard Model of Elementary Particles. Source: Wikipedia.org

The twelve fermions are the particles that make up everything tangible — stars, planets, air, oceans, and ourselves. They are a general category of subatomic particles, including fundamental particles like electrons and quarks, and composite particles like protons and neutrons, that follow the Pauli exclusion principle and constitute the building blocks of matter. Now, these fermions too have subdivisions; they are divided into two main categories: Six quarks and six leptons. Together, they are arranged in three generations, each generation containing a pair of quarks and a pair of leptons. The first generation: up quark, down quark, electron, and electron neutrino form the stable matter we encounter daily. The second and third generations consist of heavier versions that exist fleetingly in high-energy environments.

Now, what makes fermions something special is not merely their role in building matter, but a deeper quantum property: spin. Fermions possess half-integer spin values such as 1/2, 3/2, and so on, something which we discussed in quarks before. Spin, despite its name, is not literal spinning like a rotating sphere. It is an intrinsic quantum characteristic, a fundamental label that determines how a particle behaves under certain transformations.

Because of their half-integer spin, fermions obey one of the most important principles in physics: the Pauli Exclusion Principle. Now, what does the Principle state? This principle states that no two identical fermions can occupy the same quantum state simultaneously. In simple terms, identical fermions refuse to share identical conditions. This refusal is the reason matter has structure.

A very prominent way to explain this is that electrons in an atom do not collapse into the same lowest energy shell because the Pauli Exclusion Principle, also mentioned above, prevents them from doing so. Instead, they arrange themselves into distinct shells and orbitals, giving rise to the periodic table and the diversity of chemistry. Without this principle, atoms would not maintain their structure. Solid matter would not resist compression. The physical world would lose its familiar form. Knowing slightly about quarks from previous explanations, let us look at leptons first. The family of leptons, actually.

The Lepton Family

One of the primary families of fundamental particles is known as leptons. The word "lepton" comes from the Greek term *leptos*, meaning "light" or "slender," reflecting their relatively small masses compared to many other subatomic particles. Among them is a particle we already know intimately: the electron, which is the only particle found orbiting the nucleus of an atom, while the other leptons are produced in high-energy collisions, cosmic rays, and stellar fusion reactions and can exist freely in nature. Wonder why Leptons are called a family? It is because the lepton family consists of six members, arranged in three generations. Each charged lepton is paired with a nearly invisible companion known as a neutrino. Together, they form a structured pattern that mirrors the layered complexity we observe throughout nature. The first generation includes the electron and the electron neutrino — the particles that constitute everyday matter. The second and third generations contain heavier counterparts: the muon and muon neutrino, the tau and tau neutrino, making a total of six leptons. These heavier versions are rarer and short-lived, appearing mainly in high-energy environments such as cosmic rays or particle accelerators.

Starting with the electrons, perhaps the most familiar of all fundamental particles. It carries a negative electric charge and plays a central role in chemistry, electricity, magnetism, and the structure of atoms. But we know that we have come across a deeper way than that, covering topics such as quantum entanglement and its properties like coherence, decoherence and stuff.

In the obvious lepton family, beyond the electron lie its heavier cousins: the muon and the tau. The muon carries the same negative electric charge as the electron but has approximately 200 times its mass. It was discovered in 1936 in cosmic ray experiments, surprising physicists who had not predicted its existence. The tau, discovered in

1975, is even heavier — roughly 3,500 times the mass of the electron and nearly as heavy as a proton. Despite their greater mass, both particles are unstable. The muon decays in about 2.2 microseconds into an electron and neutrinos. The tau decays even more rapidly, lasting only a fraction of a trillionth of a second. These fleeting lifetimes make them rare in ordinary conditions, though they can be produced and studied in high-energy particle accelerators. So, why does nature contain heavier replicas of the electron? The Standard Model does not provide a definitive explanation. Their existence hints at a generational hierarchy embedded in the structure of matter. Perhaps these heavier generations reflect conditions present in the early universe, when temperatures and energies were vastly greater than they are today. Talking of today, modern experiments continue to examine the muon with remarkable precision. Small deviations in its magnetic properties may point toward new physics beyond the Standard Model.

Let me introduce you to a new member: A Neutrino. If electrons are familiar, neutrinos are strangers. Neutrinos carry no electric charge. They interact extraordinarily weakly with matter. Trillions of them pass through your body every second without leaving any noticeable trace. They are produced in nuclear reactions within the Sun, in radioactive decay within the Earth, and in violent cosmic events such as supernova explosions. For many years, physicists believed neutrinos were completely massless. However, experiments toward the end of the twentieth century revealed something extraordinary: neutrinos change from one type to another as they travel. This phenomenon, known as neutrino oscillation, implies that neutrinos must possess a small but non-zero mass. Because neutrinos interact so rarely, detecting them requires enormous and sensitive instruments. Massive underground detectors filled with water or ice, such as the 'IceCube' at the South Pole, are used to capture the faint signals produced when a neutrino occasionally collides with an atomic nucleus.

Despite their ghostly nature, neutrinos are abundant throughout the cosmos. During a supernova explosion, the majority of the energy released escapes in the form of neutrinos. They carry information from the deepest cores of stars, the regions that light itself cannot easily escape. If our eyes were sensitive to neutrinos, the universe would appear very different. The night sky would shimmer not only with starlight but with invisible streams of these nearly massless travellers. Neutrinos may also hold clues to one of the greatest mysteries in cosmology: why the universe contains more matter than antimatter. Certain theoretical possibilities involving neutrinos suggest mechanisms that could have tipped the balance in the early universe.

The Bigger Fishes

Looking at the heading, what do you conclude about what you are going to read? So, in this section, we shall look at the heavier fundamental particles, having finished the light leptons and starting with the second subdivision of fermions - quarks, which we have already given a light touch to. Quarks are the neighbours that stand parallel to the family of leptons. The quark concept was proposed independently in 1964 by physicists Murray Gell-Mann and George Zweig. Quarks combine in threes to form familiar particles such as protons and neutrons. Like leptons, quarks, too, are arranged in three generations, each generation containing two types, which are called *flavours*:

First generation: up and down
Second generation: charm and strange
Third generation: top and bottom.

The most familiar quarks are the up and down quarks, and they form the stable matter of our everyday world. This is the type of quark we discussed briefly; let us discuss it in depth now.

A proton consists of two up quarks and one down quark — often written as *uud*. A neutron consists of two down quarks and one up quark — written as *udd*. These combinations determine their electric charge.

An up quark carries a charge of +2/3.

A down quark carries a charge of −1/3.

Add the charges together:

For the proton:

$(+2/3) + (+2/3) + (-1/3) = +1$

For the neutron:

$(-1/3) + (-1/3) + (+2/3) = 0$

Thus, the proton has a total charge of +1, while the neutron is electrically neutral.

Quarks also possess intrinsic spin and mass. The down quark is slightly heavier than the up quark, which is a small difference with enormous consequences. If this mass difference were reversed, the proton might not be stable, and the structure of matter would change dramatically.

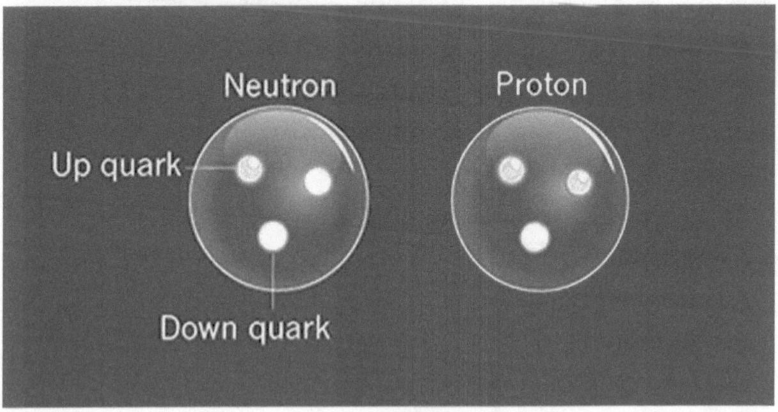

Figure 5.2: Visual Representation of the Presence of Quarks in Protons and Neutrons

Another remarkable property of quarks is something called colour charge. Despite the name, this has nothing to do with visible colour. It is a quantum property associated with the strong interaction. Quarks come in three "colours," which are conventionally labelled red, green, and blue. This naming is symbolic. It simply provides a convenient way to describe how quarks combine. To form a stable particle such as a proton or neutron, quarks must combine in a way that results in overall "colour neutrality." Just as combining red, green, and blue light produces white light, combining the three colour charges results in a neutral state. Why? Because this colour structure ensures that no two identical quarks can occupy the same quantum state, in harmony with the Pauli Exclusion Principle.

Beyond up and down lie the heavier, more exotic quarks: strange, charm, bottom, and top. These quarks are not typically found inside stable everyday matter. Instead, they appear in high-energy environments such as cosmic ray collisions or particle accelerators.

The strange quark was first identified through particles discovered in cosmic ray experiments in the late 1940s. These particles behaved oddly — they were produced quickly but decayed more slowly than expected. This behaviour led to the introduction of a new quantum property called 'strangeness', marking the beginning of a deeper understanding of quark families. The charm quark, predicted to explain certain puzzling decay patterns, was discovered in 1974 in a particle known as the J/ψ. Its discovery was so transformative that the period became known as a turning point in particle physics. The bottom quark was discovered in 1977, and finally, the top quark was discovered in 1995. The top quark is extraordinarily massive. It is so heavy that it is called the heaviest fundamental particle and decays almost immediately after being created, before it has time to bind into composite particles. If you wonder why nature requires flavours of quarks? I do too, as it remains an open question. Their structured repetition hints at a deeper organising principle, which is, perhaps, one that unifies quarks and leptons at energies far beyond what we can currently reach. Experiments in what is known as flavour physics continue to study subtle differences in how quarks transform from one type to another.

This isn't done yet. Quarks do not exist alone. They combine to form larger composite particles known as hadrons.

There are two primary categories:

Baryons — composed of three quarks (such as protons and neutrons).

Mesons — composed of one quark and one antiquark (such as pions).

In recent years, experiments have revealed even more complex combinations, including four-quark and five-quark states. These discoveries challenge earlier simplified pictures and show that quark interactions are richer than once imagined. Now that we know what hadrons are, let me tell you about one of the most fascinating, yet powerful machines made in history - The Large Hadron Collider.

Wherever the term 'particle accelerator' has been mentioned, we talk about nothing but this.

Located beneath the border of France and Switzerland, the LHC is the largest and most energetic particle accelerator ever built. It consists of a circular tunnel approximately 27 kilometres in circumference, buried deep underground. Within this vast ring, beams of protons are accelerated to velocities extremely close to the speed of light and then made to collide head-on. These high-energy collisions recreate, on a tiny scale, conditions similar to those that existed just fractions of a second after the Big Bang. In that primordial environment, quarks and gluons were not yet confined into stable protons and neutrons. By recreating such extreme conditions, physicists can study how quarks combine and how new forms of hadronic matter emerge. It is through experiments at the LHC that scientists have confirmed the existence of exotic hadrons or other unconventional arrangements.

Even now, something more to it. To get on that, ask yourself: what holds the protons and neutrons together in the nucleus of an atom? You might answer 'Strong Nuclear Force.' But, how is it produced? The answer lies in the particles called gluons, which hold quarks together. They are the carriers of the strong nuclear force — the force responsible for binding quarks tightly inside protons and neutrons. Without gluons, quarks would drift apart. Atomic nuclei would not exist. Matter, as we know it, would not hold together. The strong force is extraordinarily powerful at short distances. It overcomes the electromagnetic repulsion between positively charged protons within the nucleus. Unlike the electromagnetic force, which weakens with distance, the strong force behaves differently. As quarks attempt to separate, the force between them actually increases. This unusual behaviour leads to one of the most fascinating properties in particle physics: confinement.

Also, quarks cannot exist freely in isolation. They are permanently bound within composite particles. Nature's deep ingrafted love for quarks, it prevents the exposure of a lone quark. If one attempts to pull two quarks apart, the energy required does not isolate a single quark. Instead, that energy creates new quark–antiquark pairs, forming additional particles. The theory describing this behaviour is known as Quantum Chromodynamics (QCD), which is the quantum theory of the strong interaction. In QCD, gluons themselves carry colour charge, allowing them to interact not only with quarks but also with each other. There are eight types of gluons, each representing different ways colour charge can be exchanged between quarks. Through this exchange, quarks remain confined in colour-neutral combinations.

At extremely high temperatures, such as those that existed fractions of a second after the Big Bang, quarks and gluons may no longer remain confined. Instead, a dense state of matter where protons and neutrons "melt," allowing quarks and gluons to move freely (deconfined) rather than being bound inside hadrons, forms, known as the quark-gluon plasma. Experiments at powerful particle accelerators have recreated tiny droplets of this plasma, offering glimpses into the earliest moments of the universe.

Now, we move on to the next type of fundamental particles: the bosons. In modern physics, forces are not viewed as invisible pushes or pulls acting across empty space. Instead, they are described as exchanges of particles known as bosons, which are the messengers of interaction. If fermions are the builders of matter, bosons are the communicators. They carry influence from one particle to another, much like messengers delivering instructions between distant cities. Through their exchange, the universe maintains structure, motion, and transformation. They are of two types: Gauge Bosons and Scalar Bosons.

Gauge bosons arise from deep mathematical symmetries called gauge symmetries. These symmetries ensure that the laws of physics remain consistent under certain transformations. The photon, gluons, and W and Z bosons belong to this category. These particles have spin 1. Their existence is not arbitrary. They emerge naturally from the requirement that fundamental symmetries be preserved.

The electromagnetic force governs electric charge and magnetism. It is responsible for binding electrons to nuclei, forming atoms and molecules. It controls chemistry, electricity, magnetism, and light itself. Its carrier is the photon. Photons are massless. Because they have no mass, they can travel at the speed of light. This allows the electromagnetic force to have an effectively infinite range.

The strong force holds quarks together inside protons and neutrons. It also binds protons and neutrons together inside atomic nuclei. Its carrier is the gluon, which we have studied just now.

The weak force governs processes that change one type of particle into another. It is responsible for radioactive beta decay and plays a crucial role in the nuclear reactions that power the Sun.

Its carriers are the W and Z bosons. Unlike photons and gluons, W and Z bosons are very massive. Their masses are approximately 80 to 91 giga-electronvolts, making them far heavier than protons. Because of this large mass, the weak force operates only over extremely short distances. The heaviness of these bosons limits the range of the interaction, much like how a heavy object is harder to throw across a field. The greater the mass of the force carrier, the shorter the range of the force it mediates. Through the weak interaction, neutrons can transform into protons, emitting electrons and neutrinos. This transformation underlies radioactive decay and stellar fusion. The sunlight that warms our Earth ultimately depends on the weak force.

Now, there is a hypothetical particle that we did not include in the list, but is considered in quantum physics. Gravity governs the motion of planets, stars, and galaxies. It shapes the large-scale structure of the cosmos. In classical physics, gravity is described by Einstein's general theory of relativity as the curvature of spacetime itself. However, if gravity were to be fully described within quantum theory, it would likely require a carrier particle. That hypothetical particle is known as the graviton.

The graviton is predicted to be massless and to have a spin of 2. Yet unlike the photon, gluon, or W and Z bosons, the graviton has not been detected. Gravity remains extraordinarily difficult to reconcile with quantum mechanics. Observatories such as LIGO have detected gravitational waves, the ripples in spacetime produced by massive cosmic events, but these detections confirm the wave nature of gravity predicted by relativity, not the particle nature predicted by quantum gravity theories. The graviton remains theoretical, a reminder that our understanding of nature is still incomplete.

Distinct from the gauge bosons is a different type of boson known as a scalar boson. The most famous example is the Higgs boson, which we also call 'The God Particle.' Unlike gauge bosons, which have spin 1, the Higgs boson has spin 0. It is associated with the Higgs field, a field that permeates all of space. To understand its role, imagine space not as empty, but as filled with an invisible medium. The Higgs field is this medium. The Higgs boson is a ripple in that field. As particles move through this medium, some interact strongly with it and acquire greater mass. Others interact weakly and remain light. To understand its importance, consider a simple question: Why do particles have mass?

For many years, this was unexplained. The equations describing particle interactions required a mechanism that would allow particles to acquire mass without destroying mathematical consistency—specifically, without breaking gauge symmetries. The answer emerged through the Brout-Englert-Higgs mechanism, proposed in 1964 by Peter Higgs, François Englert, and Robert Brout (among others). It posits the Higgs field—a scalar field permeating space.

For better understanding, let us take an example: imagine a celebrity walking through a crowded room. If people cluster around, movement becomes difficult, which symbolises the gain of mass. If no one interacts, movement is effortless, symbolising lightness. Now, connect it. Particles moving through this field interact with it to varying degrees. The stronger the interaction (called the Yukawa coupling), the greater the particle's mass. Photons and gluons don't interact, remaining massless; W/Z bosons do, gaining mass via symmetry breaking.

Without the Higgs mechanism, the W and Z bosons would be massless, and the weak force would behave very differently. Atoms might not form stable structures. The universe would lack the delicate balance required for complexity.

The discovery of the Higgs boson in 2012 confirmed a crucial part of the Standard Model. It demonstrated that mass, at least for many fundamental particles, arises from interaction with a pervasive field rather than being an intrinsic fixed property.

At this point, you can take some time to grasp some breath, for we have come a very long way in understanding the fabric of matter, covering lots and lots of things, yet surplus things hide in those intricate stitches of the matter. But now, to continue the search for our answer, we will move forward towards something that is quite the opposite of what we have covered right now - Anti-Matter.

The Scissors

Hmm, why 'Scissors?' Because we studied the 'fabric' of matter right now, and anti-matter simply destroys or what we call annihilates that. Somewhat similar to what would happen when a person draws out scissors against a person drawing paper in a game of rock, paper, and scissors.

In 1928, while attempting to reconcile quantum mechanics with special relativity, Paul Dirac derived an equation describing the electron. The equation worked beautifully. But hidden within its mathematics was something unexpected: solutions corresponding to particles identical in mass to the electron, yet carrying opposite electric charge. At first, this seemed purely mathematical. Yet in 1932, Carl Anderson detected such a particle in cosmic ray experiments. It behaved like an electron but curved in the opposite direction in a magnetic field.

This particle was named the positron. It is the antimatter counterpart of the electron. From that moment onward, physics changed. Antimatter was no longer an abstract prediction. It was real.

So what actually is anti-matter? Antimatter particles share the same mass and intrinsic spin as their matter counterparts, but their charges are reversed. The electron has a negative charge; the positron has a positive charge. The proton has a positive charge; the antiproton has a negative charge. The neutron is electrically neutral, yet it too has an antimatter partner; the antineutron differs in internal quantum properties. Even quarks have antiquarks. If a quark carries a certain electric charge and colour charge, its antiquark carries the opposite electric charge and opposite colour charge. An antiproton is composed of two anti-up quarks and one anti-down quark. An antineutron contains two anti-down quarks and one anti-up quark. These antiquarks bind together through gluons just as ordinary quarks do. Even entire anti-atoms have been created in laboratories. Scientists have combined positrons and antiprotons to form anti-hydrogen atoms. These experiments allow researchers to test whether the laws of physics apply identically to matter and antimatter.

Very well. Now, what happens when anti-matter and matter meet? When a particle encounters its antiparticle, they can annihilate. Their mass converts entirely into energy, typically in the form of high-energy photons. This is not destruction in the usual sense. It is a transformation. Matter becomes pure radiation. The energy released in such annihilations is enormous compared to chemical reactions. This is why antimatter has often captured the imagination in science fiction as a powerful energy source.

But in reality, producing and storing antimatter is extraordinarily difficult. It must be confined using magnetic fields, since any contact with ordinary matter leads to immediate annihilation. This is what makes anti-matter, the most expensive substance on earth. A single gram of anti-matter costs approx. a staggering $62.5 trillion! It is because of the immense energy and complex particle accelerator, like the LHC, needed for its creation. Its storage too requires complex electromagnetic containment in a vacuum, because of its annihilative property with matter, making it incredibly difficult and costly. Only nanograms are produced annually for fundamental physics research. There is no commercial market for antimatter, making it not for sale, and the price is a theoretical estimate of production costs, as calculated by sources like NASA.

Okay, we can create antimatter in laboratories, but how was it formed in the first place? To understand antimatter's origin, we must travel back to the earliest moments after the Big Bang.

In the intensely hot, dense conditions of the early universe, energy and matter were interchangeable. High-energy photons collided, producing particle and antiparticle pairs. These pairs could also annihilate back into photons. In that primordial environment, matter and antimatter were created in nearly equal amounts. Yet today, the universe is overwhelmingly dominated by matter. Where did all the antimatter go? This is one of the deepest mysteries in cosmology.

If matter and antimatter had been produced in exactly equal amounts, they would have annihilated completely, leaving behind a universe filled only with radiation. But a tiny imbalance existed — perhaps, we can say, one extra matter particle for every billion matter–antimatter pairs. After annihilations occurred, that slight excess of matter remained. That leftover matter eventually formed galaxies, stars, planets, and life.

Why did this imbalance occur? The answer may lie in subtle violations of symmetry in the laws of physics. The tiny asymmetries may have tipped the cosmic balance, which we are not absolutely sure about.

One final discussion before entering directly to answer our question, I have mentioned the terms dark matter and dark energy quite a lot of times before, but I have not explained them. And since we are exploring matter in depth, let us cover such things too, which are not directly a part of matter, but play a great role at the cosmic level.

Galaxies rotate too quickly. If we calculate the gravitational pull based solely on visible matter, the stars, gas, and dust, galaxies should not hold together at their observed speeds. They should tear themselves apart. Yet they remain intact. Something unseen provides additional gravitational influence. This unseen component is called dark matter.

Dark matter does not emit light. It does not absorb light. It does not interact through the electromagnetic force. Its presence is inferred only through gravity. Observations of galaxy rotation curves, gravitational lensing, and large-scale cosmic structure all point to the existence of dark matter. In fact, dark matter appears to constitute about five times as much mass as ordinary matter. What is dark matter made of? It cannot be ordinary atoms, because those would interact electromagnetically. It must consist of particles that interact very weakly with light and with normal matter. Many candidates have been proposed:

Weakly interacting massive particles.

Axions.

Sterile neutrinos.

So far, they are considered theoretical, and no direct detection has been confirmed. Dark matter remains invisible, yet its gravitational influence shapes galaxies and clusters across billions of light-years. What are axions and sterile neutrinos? Axions are extremely light, electrically neutral, pseudoscalar elementary particles. Sterile (also called inert) neutrinos are hypothetical, neutral leptons that do not interact via the weak nuclear force, unlike the three known "active" neutrinos in the Standard Model.

In 1998, observations of distant supernovae revealed something astonishing: the expansion of the universe is accelerating. Gravity should slow cosmic expansion. Yet instead of decelerating, galaxies are receding from one another at increasing rates. Some unknown component of the universe appears to be driving this acceleration. This component has been named dark energy. Unlike dark matter, which pulls through gravity, dark energy seems to act as a repulsive influence embedded in space itself.

It is estimated to constitute nearly seventy percent of the total energy content of the universe. One possibility is that dark energy corresponds to a property of the vacuum — a small but nonzero energy density inherent to empty space. Another possibility is that it reflects a dynamic field evolving over cosmic time.

True Vacuum

If you might or might not be fierce on me for making you read so much just for a simple answer, but I suggest you not to be, for it won't go to waste.

Let us start now. The True Vacuum When we hear the word vacuum, we instinctively imagine emptiness — a perfect void, a region devoid of matter, devoid of light, devoid of activity. Classical physics encouraged this intuition. Remove all particles from a box, pump out every molecule of air, block all radiation, and you are left with nothing.

But quantum physics does not permit such simplicity.

In the quantum world, emptiness is never truly empty.

The concept of a true vacuum goes far beyond the mere absence of particles. It is not a silent void but the lowest possible energy state of all quantum fields. It is what remains when every particle has been removed, when every excitation has settled, when the universe locally reaches its ground state.

What actually happens is that an electron and a positron annihilate. Their mass converts into energy, typically photons. The particles vanish as individual identities. What remains is radiation. If radiation escapes and energy disperses, what lingers in the region where the particles once existed? Not a classical void. What remains is the quantum field in its ground state — the vacuum.

In this sense, a vacuum can be viewed not merely as emptiness but as the aftermath of annihilation. When matter and antimatter neutralise one another, they return the field to its baseline condition. The excitations disappear, but the underlying field persists.

Imagine the surface of a calm lake. Throw two equal but opposite ripples into it. When they cancel perfectly, the surface becomes flat again. Yet the water remains. The medium remains. Only the disturbance has vanished. The vacuum is that calm surface — not the absence of water, but the absence of waves.

The Quantum field theory gives us the explanation for this. Quantum field theory teaches us that every particle is an excitation of an underlying field.

The electron field gives rise to electrons.

The quark fields give rise to quarks.

The electromagnetic field gives rise to photons.

Even when no particles are present, these fields permeate all of space.

The lowest energy configuration of these fields is called the vacuum state. But the uncertainty principle forbids absolute stillness. Energy and time cannot both be defined with perfect precision. As a result, tiny fluctuations constantly ripple through the vacuum. For extremely brief intervals, particle and antiparticle pairs can momentarily appear and disappear. These are often described as virtual particles. They are not directly observable in the way stable particles are, but their effects are measurable. They slightly modify atomic energy levels. They contribute to forces between objects separated by tiny distances. They shape the behaviour of black holes.

Something Out Of Nothing

We all know that a black hole forms when massive matter collapses under its own gravity. Its gravitational field becomes so intense that beyond a certain boundary, called the event horizon, nothing can escape, not even light. Near this boundary, the quantum vacuum behaves in remarkable ways.

As described earlier, particle and antiparticle pairs constantly flicker into existence in the vacuum. Normally, these pairs annihilate almost immediately, leaving no lasting trace. But near a black hole, gravity can intervene. If such a pair forms close to the event horizon, it is possible for one particle to fall into the black hole while the other escapes to infinity. The pair is separated.

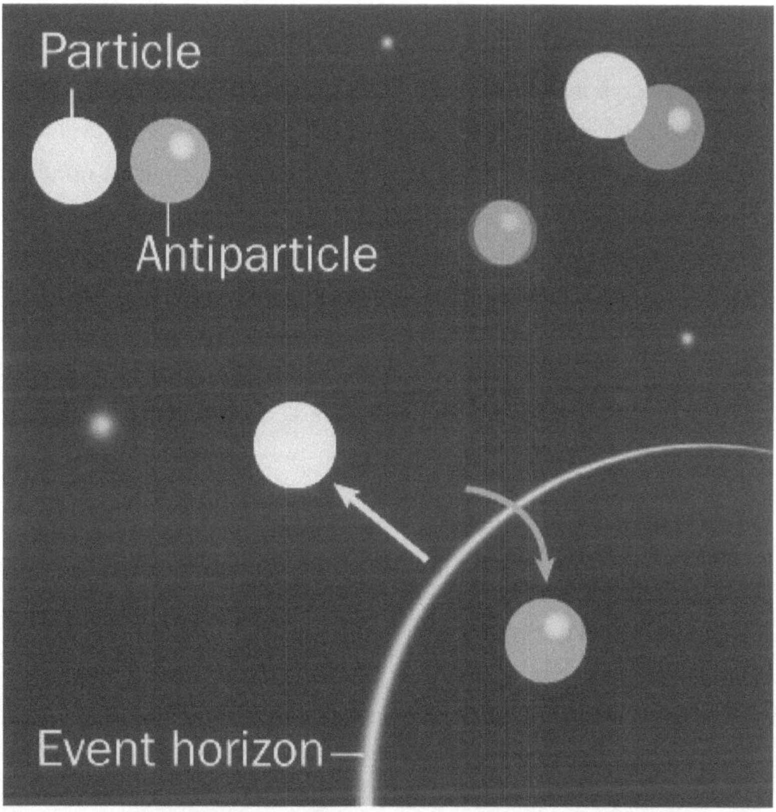

Figure 5.3: Particle & Antiparticle Separation by Black Holes. Source: Sciencenews.org Black Holes in the Bathtub

To an outside observer, it appears as though the black hole has emitted a particle. This phenomenon is known as Hawking radiation, proposed by Stephen Hawking in 1974. Through Hawking radiation, the vacuum reveals its generative power.

But the black holes pay a price for it, as a crucial question remains. If one particle escapes, where does its energy come from? Because the energy conservation must hold.

The answer is subtle. The particle that falls into the black hole can carry negative energy relative to the outside universe. When this negative energy is absorbed, the black hole's total mass decreases. In effect, the black hole pays the energy cost.

The escaping particle carries positive energy away, while the infalling partner reduces the black hole's mass. Over immense timescales, this process causes black holes to lose mass gradually. They evaporate. The rate of evaporation depends on the black hole's mass. Smaller black holes radiate more intensely. Larger ones radiate extremely slowly. For stellar mass black holes, the evaporation timescale vastly exceeds the current age of the universe. But in principle, no black hole is eternal.

Denouement V

We began this journey with a simple curiosity.

What lies beneath the matter?

What is hidden inside the atom, inside the proton, inside the quark?

And what do we mean when we speak of emptiness?

Step by step, the fabric unfolded before us. We explored the architecture of matter, encountered antimatter, and looked at something matter-like yet not matter - the dark matter. And finally, we arrived at the vacuum, which once seemed the simplest concept of all. We learned that a vacuum is not emptiness. It is not a hollow void carved out of nothingness. It is the ground state of quantum fields, restless with fluctuations. Particle–antiparticle pairs flicker in and out of existence. Even black holes draw upon this subtle sea, emitting radiation as quantum effects separate pairs at their horizons.

But, before we close this chapter, let us pause at one recurring idea.

Many times, we have mentioned that mass is converted into energy.

Annihilation releases energy.

Fusion releases energy.

Even black holes lose mass through radiation.

But how is this possible?

How does something with mass transform into something as intangible as energy? Not only this, but also our simplest measurements on which the foundations of physics stand, do we really measure them precisely, or ignore some minute side actors?

We have examined matter, antimatter, forces, vacuum, and cosmic structure. Now we must examine the foundations of measurement itself — the relationship between mass and energy, the reliability of observation, the profound equations that tie them together, and many more such great things in our next chapter...

6

The Meticulous Way

"How shocking that as your velocity increases, your mass grows, yet you never truly gain an ounce."

Suppose you are given a simple task. "Bring one kilogram of vegetables from the market." You walk through the market, choose the veggies carefully, weigh them on the scale, pay the vendor, and head toward your home. But imagine this. Halfway home, the vegetables suddenly disappear. Neither dropped, nor stolen, nor misplaced. Vanished. One full kilogram of mass vanishes from existence. You would panic. What explanation could you possibly give? How do you explain that one kilogram of matter ceased to exist?

But let me relieve you of one concern. If one kilogram of mass were truly "destroyed," there would be no house left to return to.

No street. No market. No city. Why? Because one kilogram of mass, if converted entirely into energy, would release an amount of energy so enormous that it would rival the most powerful explosions humanity has ever conceived. It sounds like science fiction and hypothetical, but it's not a mere imagination. For comparison, converting 1 kg of mass into pure energy is equivalent to approximately 1,400 to 1,500 Hiroshima-sized bombs. Before we rush toward it, however, let us deepen the mystery, as there is one more statement that appears equally unsettling.

The above given quote states that mass increases with velocity. So, what does that mean? If you run, do you become heavier? If a spaceship travels faster, does its mass grow? And if mass can turn into energy, what truly is mass? Let me tell you that these are not casual curiosities, because we often assume that when we measure something, the result reflects a fixed property of reality. Like, one meter is one meter and one second is a second.

Okay, we have a lot of curious questions, but what is meticulous in it? The idea that mass can transform into energy, and that mass depends on velocity, did not emerge from imagination. It emerged from meticulous reasoning, careful experiments, and logical consistency. To know the answer to all these questions, let us proceed very carefully in our chapter.

Cosmic Currency

So, let us begin with nothing but our everyday life. In our everyday lives, we often treat mass as something simple and obvious. A kilogram of apples weighs a kilogram whether we are in a kitchen, a laboratory, or a marketplace. A stone feels heavy in the hand; a feather feels light. The distinction seems straightforward: objects possess a certain quantity of matter, and that quantity determines their mass. But the deeper we travel into modern physics, the more this simple picture begins to dissolve.

Mass, the very property that appears to define matter, which we study as the amount of matter packed in a body, turns out to be something far more subtle. In fact, when physicists began to explore the fundamental structure of matter — the world of quarks, electrons, and the forces binding them — they discovered that the majority of what we call mass is not matter at all. Instead, it is energy.

This was the realisation that did not come overnight. It emerged slowly through decades of experiments, theories, and deep reflections on how nature operates at its most fundamental level. At the centre of this realisation lies one of the most famous insights in the history of science: the profound relationship between mass and energy, the revolutionary
'Mass Energy Equivalence Formula' by Albert Einstein.
$E = mc^2$.

Yet, as we will see, this equation is often misunderstood. It does not merely tell us that mass can turn into energy. It reveals something even more astonishing, that mass itself is a form of energy. To understand what this truly means, we must begin by examining the structure of the matter that makes up our bodies. Meanwhile, keep this formula in your head for its utmost great importance in this and other chapters, too.

You might imagine that your mass is simply the sum of the masses of the particles that compose you — the electrons, protons, and neutrons inside every atom. After all, if you add up the masses of all these particles, shouldn't you obtain the mass of your body? You might think that the answer is yes, but surprisingly, the answer is a big no. When physicists performed the calculations carefully, they found something remarkable. The masses of the fundamental particles, namely the quarks and electrons, account for only a tiny fraction of the total mass of ordinary matter. In fact, they contribute only about 1 to 5% of the mass of the human body. Take a pause for a moment and let this settle. If the fundamental particles themselves contribute only a few percent, then where does the remaining 95% come from? Shockingly, the answer lies not in the particles themselves but in the energy that binds them together.

To know more about that energy, we need to look at matter closely. I know you're used to it, but let us zoom in again on the fabric of matter, to check what really happens. Now, here's something which we already know and something which we don't - Inside every proton and neutron are quarks, bound together by the strong nuclear force. This force is carried by particles called gluons, and it is extraordinarily powerful. The quarks inside a proton are not quietly sitting next to each other like marbles in a box. Instead, they are constantly moving, interacting, and exchanging gluons in a restless dance governed by the rules of quantum chromodynamics. This motion and interaction carry enormous amounts of energy. When physicists calculate the total energy stored in this dynamic system, which is the motion of quarks, the gluon fields, and the interactions between them, they discover that this energy is immense.

Now comes the remarkable step. Using Einstein's insight, if we divide that energy by the square of the speed of light (c^2), the result appears as mass. Not just a random number, but the exact 95% of the mass that was missing. In other words, the majority of the proton's mass does not come from the masses of the quarks themselves. It comes from the energy of their interactions.

At this point, we begin to realise that something profound has happened to our understanding of mass. Traditionally, mass was thought of as a fundamental property of matter — something intrinsic and permanent. But modern physics tells a different story. Instead of treating mass as an independent concept, physicists often view it as a particular way of measuring energy. More precisely, what we call mass is often expressed as

$m = E / c^2$

This means that the quantity we measure as mass is simply energy divided by the square of the speed of light. From this perspective, the concept of mass begins to blur. Rather than thinking of mass and energy as separate entities, we begin to see them as two different ways of describing the same underlying physical reality. Energy becomes the true currency of the universe. Mass is merely one of the ways that currency appears.

But, if mass is somewhat energy only, then what about the remaining 1–5% of mass that does come from particles themselves? Where does that originate? For this answer, we must turn to many times mentioned and one of the most fascinating discoveries in modern particle physics: the Higgs field. According to the Standard Model of particle physics, space is permeated by an invisible field known as the Higgs field. Particles moving through this field interact with it, and through this interaction, they acquire energy. When quarks and electrons interact with the Higgs field, they gain energy from that interaction. This energy appears as their 'rest mass.' This energy manifests itself as what we perceive as mass. And once again, energy lies at the root.

By now, a pattern should be emerging. Mass is not a fundamental substance. Instead, it is a form of stored energy. When energy becomes confined within systems, like inside protons, inside atoms, inside fields, then it can appear as mass. This idea leads to some fascinating consequences. For example, if the internal energy of a system increases, then according to Einstein's relation, its mass should also increase. This might sound strange at first, but it follows directly from the principle that energy and mass are equivalent.

To get a better view of this, let us take an example of boiling water. Consider a pot of water on a stove. As the water heats up, the molecules inside begin to move faster and faster. Their vibrations grow more energetic as the temperature rises. This increased motion represents additional internal energy. And according to Einstein's relation, more energy means slightly more mass. So yes, when water is heated, its mass increases. The increase is extremely tiny, far beyond what an ordinary scale could measure, but in principle it is real. In the language of physics, the additional thermal energy contributes to the total energy of the system, and therefore to its mass. Thus, even something as ordinary as boiling water demonstrates the deep connection between mass and energy. Something extremely important to note here is that we are taking the 'rest mass' into consideration.

At this point, a very simple question arises. If increasing internal energy increases mass, then what about motion? Okay, suppose we take a ball and throw it across a field. While it is moving, it clearly possesses kinetic energy. Should that kinetic energy also increase the ball's mass?

No, it won't. But you might argue on the fact that I have mentioned that mass increases with velocity. For now, be patient and assume that it doesn't, for the reasons I'll provide.

When we heat water, we add energy inside the system. The molecules vibrate more vigorously, increasing the internal energy of the system itself. Because this energy becomes part of the system, it contributes to its mass. But when we throw a ball, the situation is different. The kinetic energy associated with the ball's motion depends on the frame of reference — the perspective from which we observe the motion. To someone standing on the ground, the ball has kinetic energy. But to someone moving alongside the ball at the same speed, the ball appears stationary. From that observer's perspective, the kinetic energy disappears. Because kinetic energy depends on the observer, it does not represent a permanent change in the intrinsic mass of the object. Let's bring measurement into this.

Imagine trying to weigh a moving ball. To measure its weight accurately, a weighing scale must move with the ball so that the ball rests on the scale without relative motion. But once the scale moves alongside the ball, the ball is effectively stationary relative to the scale. In that frame of reference, the kinetic energy associated with the motion disappears. Thus, the measurement reveals the same mass as if the ball were at rest. For this reason, physicists today prefer to speak of rest mass, which remains constant regardless of motion.

The correct way to express the energy of a system is to recognise that it contains multiple contributions.

The total energy can be written conceptually as:

$$E_{(total)} = E_{(rest)} + E_{(motion)}$$

Here:

$E_{(total)}$ - total energy of the system

$E_{(rest)}$ - energy present even when the object is stationary

$E_{(motion)}$ - energy associated with motion

The rest energy is given by Einstein's famous relation

$$E_{(rest)} = mc^2$$

This rest energy represents the intrinsic energy contained within the system — the energy due to particle masses, interactions, and internal structure, and not the total energy.

Motion energy, on the other hand, depends on how the object moves relative to an observer. Thus, the total energy of a moving object includes both contributions, but the mass itself corresponds only to the rest energy.

This brings us back to the famous equation. Many people interpret $E = mc^2$ as saying that mass can turn into energy, as happens in nuclear reactions. While that is certainly true, the equation's deeper message is far more profound. It tells us that mass already is energy. Every object around us, be it every stone, every tree, every human body, contains an enormous amount of energy simply by virtue of existing.

One way to think about mass is to imagine it as a kind of cosmic accounting system.

Energy flows through the universe in many forms — motion, radiation, heat, fields, and interactions. Mass represents the portion of energy that has become localised and stored within matter. Thus, when physicists weigh an object, they are not simply measuring "how much matter" it contains. They are measuring the total energy locked inside its structure. Thereby, I like to call energy the 'Cosmic Currency'.

Given this fundamental nature of mass, can we apply it to photons and try to measure their mass, as in the moving ball example?

Superb Photons

In the previous segment, we explored the profound relationship between mass and energy, which is the idea that mass is simply another form of energy, expressed in the famous mass-energy equivalence formula. This equation becomes especially intriguing when we apply it to photons. Photons clearly carry energy. Sunlight warms the Earth. Laser beams can cut metal. Solar panels convert light into electricity. These phenomena all demonstrate that photons contain energy. So a natural question arises:

If photons have energy, can we calculate their mass using the energy–mass relation? At first glance, it seems we should be able to. But the answer is no. Why? Because it is one of the most fundamental properties of light, that photons can never be at rest.

To understand why photons cannot be at rest, we must first understand something remarkable about the speed of light. In everyday life, speeds are relative. If you are sitting in a moving train and throw a ball forward, someone standing outside the train sees the ball moving faster than you do. This is because velocities normally add together. If the train moves at 50 km/h and you throw the ball forward at 10 km/h relative to the train, an observer on the ground sees the ball moving at 60 km/h. Suppose you could somehow chase a beam of light. Imagine building a spaceship capable of travelling at enormous speeds. You accelerate closer and closer to the speed of light. At last, you approach the speed of the photon you are chasing. What would you expect to see? Intuition suggests that the photon should appear to slow down relative to you, just as a car appears slower when you drive alongside it. For centuries, physicists assumed that light behaved the same way. But experiments in the late nineteenth century revealed something astonishing: light does not obey this rule. No matter how fast the observer moves, the speed of light remains the same. If that train is moving at 0.5c and you turn on a flashlight, both you and the person on the ground must measure the light at exactly 1.0c. This strange behaviour was explained in the theory of relativity.

And, according to relativity, the speed of light in a vacuum is denoted by the constant 'c', which is the same for all observers. Its value is approximately

$c = 3 \times 10^8$ meters per second.

Nothing in the universe can travel faster than this speed.

Now we reach an important point, which is why mass can't be calculated just because photons are never at rest. If photons always travel at the speed of light, then they can never be stationary. See, there is no frame of reference in which a photon is at rest. Every particle with mass can be brought to rest. A ball can sit on a table. A proton can exist inside an atom without moving through space at relativistic speeds. But photons are fundamentally different. Their very nature requires them to move at the speed of light. This means that photons have no rest frame. And since they can never be at rest, there is no way to measure a rest mass for them.

Because they don't have any rest mass, the formula $E = mc^2$ cannot be applied to them, as the formula only serves to find out the energy of the rest mass of any body.

Now, if photons have zero mass, yet they clearly possess energy, does that contradict Einstein's famous equation? The answer is no. Instead, photons obey a different energy relationship derived from quantum theory:

$E = \hbar\nu$

where:

$E \rightarrow$ energy of the photon

$\hbar \rightarrow$ Planck's constant

$\nu \rightarrow$ frequency of the light wave.

This equation shows that the energy of a photon depends on the frequency of the light. Blue light, with a higher frequency, carries more energy than red light. But in all cases, the rest mass remains zero. Thus, photons are described as massless particles.

However, physicists once considered another way of thinking about mass. Which we consider too, for now. Instead of defining mass using rest energy, we could define it using total energy. Suppose we write the relation

$m = E_{(total)} / c^2$ or $m = h\nu / c^2$

Here, instead of using rest energy, we use the total energy of the particle. This includes every form of energy it possesses. Using this definition, even a photon, which clearly has energy, would appear to have a certain mass. This concept became known as relativistic mass. Under this interpretation, the faster an object moves, the more energy it has, and therefore the greater its relativistic mass. Using this idea, we could calculate a finite mass for photons based on their energy. But wait a minute, wasn't this analogy wrong? Because $E = mc^2$ works only on rest mass? Actually, this analogy wasn't completely wrong; instead, it was abandoned. So why did physicists abandon this interpretation? To understand the reason, we must examine what happens when objects approach the speed of light.

One of the most important predictions of relativity is that accelerating an object becomes increasingly difficult as its speed approaches the speed of light. The closer an object moves toward c, the more energy is required to increase its speed further. Mathematically, the required energy grows rapidly and eventually approaches infinity. This means that no object with non-zero rest mass can ever reach the speed of light. No matter how powerful the engine, no matter how advanced the technology, accelerating a massive object to light speed would require infinite energy.

Now consider what happens if we try to apply the relativistic mass definition to photons. Using the equation

$$m = E_{(total)} / c^2$$

we might attempt to calculate the photon's mass. But here we encounter a problem. If a particle with rest mass attempts to reach the speed of light, the energy required becomes infinite. If the energy becomes infinite, then the calculated mass — using the above relation — also becomes infinite. This leads to an absurd result. It would imply that photons possess infinite mass. But this cannot be correct, because photons clearly behave as particles with finite energy. Thus, the relativistic mass interpretation leads to contradictions.

But there is a way to clear these contradictions. Physicists realised that there were two ways to define mass, each leading to different conclusions.

Case (i): Particles With Rest Mass -

If a particle has a non-zero rest mass

$$M_{(rest)} \neq 0$$

then as its velocity approaches the speed of light

$$M_{(moving)} = \infty$$

This reflects the fact that accelerating such a particle to light speed requires infinite energy.

Case (ii): Particles With Zero Rest Mass -

If the particle has zero rest mass

$$M_{(rest)} = 0$$

then it can move at the speed of light without requiring infinite energy. Its energy remains finite, even though its speed equals 'c'.

So why did physicists abandon this idea? Although the relativistic mass concept was mathematically possible, it created unnecessary complications.

Physicists had to constantly distinguish between rest mass and moving mass. The equations became messy, and the interpretation of physical quantities became confusing. Modern physics, therefore, adopted a simpler approach:

Mass is defined strictly as rest mass.

Objects with rest mass cannot reach the speed of light.

Objects with zero rest mass must always travel at the speed of light. This definition removes ambiguity and keeps the equations of relativity elegant and consistent. Thus, photons are described simply as massless particles that carry energy and momentum while travelling at the speed of light.

Photons, therefore, occupy a very special place in the universe.

They are neither ordinary particles nor traditional waves.

They have no rest mass.

They cannot slow down.

They cannot stop.

They are pure packets of energy moving eternally at the universe's ultimate speed. Yet despite having no mass, photons still possess momentum and can exert pressure. Sunlight pushes on solar sails. Laser beams can manipulate microscopic particles. Radiation from stars can drive gas away from stellar surfaces. All of these effects arise from the momentum carried by mere massless photons. It is for all these reasons that photons deserve the title "Superb Photons."

Boom!

Now that we have explored the deeper meaning of mass and its intimate relationship with energy, the idea that mass is a form of stored energy should be becoming clearer. With this understanding, we can return to the simple scenario introduced at the beginning of this chapter — the seemingly ordinary one kilogram of vegetables. According to the mass–energy relation, the energy contained in a given mass is expressed through the equation-

$$E=mc^2$$

If we substitute a mass of 1 kilogram into the equation, we obtain

$$E = (1\text{ kg}) \times (3\times10^8)^2$$

$$E = 9 \times 10^{16}\text{ joules}$$

This number is difficult to grasp directly. It represents ninety quadrillion joules of energy, an amount so immense that ordinary experience provides little guidance for understanding it. To make sense of such a number, physicists often compare it with events whose energies are better known. One such comparison comes from the most destructive devices humans have ever created: nuclear weapons.

Earlier, we mentioned that the energy contained in one kilogram of mass is roughly comparable to the energy released by 1400–1500 atomic bombs similar to the one dropped on Hiroshima. Let us now see how that comparison arises. The bomb dropped on Hiroshima during the Second World War was known as Little Boy. It was approximately three meters long and weighed about 4,400 kilograms. Inside this massive device was about 64 kilograms of enriched uranium-235, the isotope capable of sustaining a nuclear chain reaction.

However, the bomb was highly inefficient by modern standards.

Only a small portion of the uranium actually participated in the nuclear reaction. In fact, only about one kilogram of the uranium underwent nuclear fission during the explosion. Even within that reacting kilogram, not all of the mass was converted into energy. In nuclear fission, the energy released comes from a slight difference between the mass of the original nucleus and the combined masses of the fragments produced after it splits. This difference in mass, called the mass defect, is transformed into energy according to Einstein's relation. In the Hiroshima bomb, the amount of matter actually converted into pure energy was extremely small: roughly 0.6 to 0.7 grams. Let us see what that means quantitatively.

If we take approximately 0.7 grams, or 0.0007 kilograms, and apply the same equation

$$E = mc^2$$

we obtain

$$E = 0.0007 \times (3\times10^8)^2$$

$$E \approx 6.3 \times 10^{13} \text{ joules}$$

This corresponds to about 63 trillion joules of energy. Despite being only a tiny fraction of a kilogram, this energy release was enough to devastate an entire city. The explosion destroyed much of Hiroshima, caused massive fires, and released radiation that affected survivors for decades.

It is important to understand that the bomb did not literally turn one kilogram of uranium into energy all at once. Instead, the fission process rearranged atomic nuclei in such a way that a small difference in mass appeared between the starting material and the resulting fragments. That tiny difference, that is even less than a gram, was what became energy.

So, how did this transformation occur?

The mechanism involves a chain reaction. When a uranium-235 nucleus absorbs a neutron, it becomes unstable and splits into smaller nuclei. During this process, additional neutrons are released, which then strike other uranium nuclei, causing them to split as well. If enough uranium is packed together, reaching what is known as critical mass, the reaction grows explosively. In the Hiroshima bomb, conventional explosives were used to rapidly assemble pieces of uranium into a supercritical configuration. When the reaction began, billions upon billions of atomic nuclei split in a fraction of a second, releasing enormous energy.

The details of nuclear weapon design are complex and involve sophisticated physics and engineering, far beyond the scope of our discussion here. For our purposes, the essential point is simple: even a very small loss of mass can produce an immense amount of energy.

Now let us return to our earlier calculation. If 0.7 grams of mass produced the energy of the Hiroshima explosion, then how many such explosions would equal the energy contained in one kilogram of mass? To find this, we compare the two values.

One kilogram corresponds to

9×10^{16} joules

while the Hiroshima explosion corresponds to roughly

6×10^{13} joules

Dividing these values gives

$9 \times 10^{16} / 6 \times 10^{13} \approx 1500$

This means that one kilogram of mass converted completely into energy would release about fifteen hundred times the energy of the Hiroshima bomb. This is why I earlier said that if your one kilogram of vegetables were somehow converted entirely into energy, the result would be unimaginably catastrophic. The energy released would be comparable to the combined explosion of well over a thousand atomic bombs.

Another atomic bomb was dropped a few days later on the Japanese city of Nagasaki. This weapon, known as Fat Man, used a different nuclear material: plutonium-239. The design of this bomb was also different. Instead of assembling pieces of nuclear material together, powerful explosives compressed a plutonium core inward from all directions. This method, known as implosion, created the extremely high densities required to sustain the chain reaction. The Nagasaki bomb contained about 6.2 kilograms of plutonium-239, much less nuclear material than the Hiroshima bomb. Yet its design allowed the nuclear reaction to proceed more efficiently. During the explosion, approximately one gram of mass was converted into energy.

Using Einstein's relation once again,

$E = (0.001\ \text{kg}) \times (3\times10^8)^2$

$E = 9 \times 10^{13}$ joules

This is roughly 90 trillion joules, slightly greater than the energy released in the Hiroshima explosion. The improved efficiency of the implosion design allowed a larger fraction of the nuclear material to participate in the reaction. Although the Nagasaki bomb contained less fissile material overall, the portion that actually converted mass into energy was larger. In that sense, the device was significantly more efficient. The devastation caused by both explosions remains a sobering reminder of the enormous energy stored within matter. What appears to us as solid mass contains vast reservoirs of energy locked inside atomic nuclei.

Relativity

Velocity is one of the simplest ideas we encounter in physics. Even a child intuitively understands it. If a car travels from one city to another, we can measure how fast it moves. If a runner crosses a track, we record their speed. Velocity simply tells us how quickly something changes its position over time.

Mathematically, it is defined as:

$v = \text{displacement}/\text{time}$

This idea seems so straightforward that for centuries, physicists assumed velocity behaved in ways that matched everyday intuition. If one object moved faster than another, it simply meant it covered more distance in the same time. Motion appeared to be absolute and uncomplicated. But as scientists began exploring the behaviour of light and extremely fast-moving particles, something astonishing emerged.

You might or might not know, velocity, when pushed toward extreme limits, begins to reshape the very structure of reality. Distances shrink. Time flows differently. Events that appear simultaneous for one observer cease to be simultaneous for another. All of these effects arise from a single fundamental fact about nature, something which we discussed earlier: The speed of light in vacuum is constant for all observers. And if this is true, then space and time themselves must adjust to preserve that constant speed. But, before we dive deeper, I noticed that the word 'relativity' has been used 14 times in the book, and since we are about to discuss relativity, I thought of mentioning this fact.

So, Sir Albert Einstein, in 1905, formed the foundation of the Theory of Special Relativity, which developed the above-mentioned insights. The theory is built upon two fundamental principles.

The first principle states that the laws of physics are the same in all inertial frames of reference. An inertial frame is simply a frame that is moving with constant velocity and not accelerating. This means that if you are sitting inside a smoothly moving train with no windows, you cannot perform any experiment that would reveal whether the train is moving or standing still. The laws of physics behave exactly the same.

The second principle states that the speed of light in a vacuum is always the same constant value, represented by the symbol 'c'. This is true for every observer, regardless of how fast they are moving relative to the light source.

These two simple statements lead to consequences that completely contradict our everyday intuition. To preserve the constant speed of light, the universe must behave in a very unusual way. Objects appear shorter. Mass behaves strangely. Clocks slow down. And the idea of a universal present moment begins to dissolve. All of these extraordinary effects arise from the principles of Special Relativity.

After publishing Special Relativity in 1905, Einstein quickly realised that something important was still missing. His theory explained how space and time behave at high speeds, but it did not include gravity. At that time, gravity was still described using Newton's famous law of gravitation. According to Newton, every mass attracts every other mass through a force acting across space. This theory worked extremely well for centuries. It explained the motion of planets, comets, and falling objects with remarkable accuracy. But Einstein noticed a troubling inconsistency. Newton's theory implied that gravitational effects travel instantaneously across space. If the Sun were to suddenly move, Earth would immediately feel the change in gravitational force. But don't you think it is absurd, as Special Relativity had already established that no information can travel faster than light, so gravity could not possibly act instantaneously.

After nearly a decade of intense work, Einstein finally presented his new theory in 1915. It became known as the General Theory of Relativity.

In this theory, gravity is no longer treated as a force pulling objects together. Instead, gravity is described as the curvature of space and time caused by mass and energy, which we have already mentioned often before. Imagine placing a heavy bowling ball on a stretched rubber sheet. The sheet bends under the weight of the ball. If you roll a smaller marble across the sheet, the marble will follow curved paths around the ball. Similarly, massive objects such as stars and planets bend the fabric of space-time around them. Other objects simply follow the curved paths created by this distortion. Thus, planets orbit stars not because they are pulled by an invisible force, but because they are moving through curved space-time.

Einstein's theory made several predictions that sounded almost unbelievable at the time. Over the past century, however, each of these predictions has been confirmed through experiments and observations. Some of the most remarkable predictions include:

- Gravitational time dilation, where time flows more slowly near massive objects.
- Gravitational lensing, where light bends around stars and galaxies.
- Black holes are regions where gravity becomes so strong that nothing can escape.
- Gravitational waves are ripples in space-time produced by massive accelerating objects.

Now, you might think, how do these theories differ? Although Special Relativity and General Relativity are deeply connected, they describe different physical situations. Special Relativity focuses on motion at constant velocity and the strange behaviour that emerges when objects approach the speed of light. General Relativity extends these ideas further, incorporating gravity and acceleration into the picture.

If you are still confused, then in simple terms:

Special Relativity explains how velocity affects space and time.

General Relativity explains how mass and gravity shape space and time.

One deals primarily with motion, while the other deals with the structure of the universe itself. You can think of Special Relativity as a special case within the broader framework of General Relativity.

Now, let me tell you something. Relativity as a whole is a very vast topic, so explaining all of it can be our story for some other day. But for now, our focus will remain mainly on Special Relativity because we are interested in understanding the astonishing consequences that arise when objects move at extremely high velocities. As we noted earlier, a very curious question is: if you run, do you become heavier? And what is meticulous in it?

So, let's start the search for our answer in the theory of Special Relativity and explore some of its other postulates too. From Einstein's theory emerge several extraordinary consequences:

Length contraction
Time dilation
Increase of relativistic mass
Relativity of simultaneity
Relativistic Doppler shift

Each of these effects sounds strange at first. Some even seem impossible. But they have all been experimentally verified many times. We will cover each of these in an ordered manner.

The first consequence we face is length contraction. Now, to understand it, imagine observing a spaceship flying past you at an extremely high speed. To the passengers inside the ship, everything appears normal. The length of the ship seems unchanged. Their rulers measure the same distances they always did. But to you, who is standing outside and watching the ship rush past — something strange happens. The spaceship appears shorter in the direction of motion, neither compressed due to physical forces nor squeezed by pressure. But fundamentally shorter in space itself. This phenomenon is known as length contraction. It is described mathematically by the Lorentz transformation, discovered by Hendrik Lorentz. The equation for length contraction is:

$$L = L_0\sqrt{1 - (v^2 / c^2)}$$

Where:

L = observed length
L_0 = proper length (length measured at rest)
v = velocity of the moving object
c = speed of light

The factor

$$\sqrt{1 - v^2 / c^2}$$

is known as the Lorentz factor component.

As velocity increases, this factor becomes smaller, causing the observed length to shrink. It is pretty easy to calculate, too. Let us take a look at

some examples:
If an object moves at 10% of the speed of light, the contraction is extremely tiny, far too small to notice.
At 50% of the speed of light, the object shrinks to about 87% of its original length.
At 90% of the speed of light, the length reduces to about 44%.
At 99% of the speed of light, the object shrinks to only 14% of its original length.
And if an object could reach the speed of light itself? The equation predicts something remarkable. The length becomes zero. You might say, how can it reach the speed of light. Of course, no object with mass can reach the speed of light, so this extreme situation never actually occurs. But the equation reveals how dramatically space itself changes at relativistic speeds. To someone moving with the object, nothing changes. To an outside observer, however, space itself appears compressed. Different observers can measure different lengths for the same object depending on their relative motion. Why does this occur? We'll answer this at last after covering all of the postulates, among which we have already discussed time dilation.

So, just as velocity changes length, it also changes another fundamental property of matter: mass. Yes, we're getting there. In classical physics, mass is constant. A rock remains the same mass whether it is sitting on the ground or moving through space. But relativity tells a different story. As an object's velocity increases, its resistance to acceleration also increases. It becomes harder and harder to push the object faster. This effect is sometimes described using the concept of relativistic mass.

The equation describing this relationship is:

$$m = m_0 / \sqrt{1 - v^2 / c^2} \text{ or } m = \gamma m_0$$

Where:

m = relativistic mass

m_0 = rest mass
v = velocity
c = speed of light
γ = Lorentz factor
Notice two important things here. The denominator is the same Lorentz factor we saw earlier. As velocity approaches the speed of light, the term

$1 - v^2 / c^2$

approaches zero. And when a denominator approaches zero, the value of the fraction becomes infinite. This means relativistic mass becomes infinite. The second important thing to notice is - this is the exact phenomenon we mentioned earlier in Cases (i) and (ii). So, do I mean that moving mass and relativistic mass are the same? Yes, absolutely. At speeds close to light speed, even a tiny particle begins behaving as if it were enormously massive. If an object could reach exactly the speed of light, the equation predicts something extraordinary. Its mass would become infinite, as said. And accelerating something with infinite mass would require infinite energy. This is why objects with rest mass can never reach the speed of light. Nature itself prevents it. Now we know why, when we move, we gain mass but never truly gain an ounce. It does not mean that the object physically accumulates infinite matter. Instead, it means that the energy required to continue accelerating the object becomes infinitely large.
Now, something new is that Modern physics usually avoids the term relativistic mass and instead describes the effect in terms of energy and momentum, but the underlying idea remains the same: The faster something moves, the harder it becomes to accelerate it further.

The desynchronisation of "Now", perhaps one of the most surprising consequences of relativity, is also known as the relativity of simultaneity. What is it? Again, in everyday life, we assume that if two events occur at the same time for one observer, they must occur at the same time for everyone. This assumption feels so obvious that it rarely occurs to us that it could be wrong. Yet according to Special Relativity, simultaneity is not absolute.

Two observers moving relative to one another may disagree about whether two events occurred at the same moment. This idea is best illustrated through a famous thought experiment often called the train and lightning experiment, originally described by Albert Einstein.

Imagine a long train moving at a constant velocity along a straight track. At the exact moment the centre of the train passes a stationary observer on the platform, two lightning bolts strike the front and rear ends of the train. To the observer standing on the platform, the flashes of light from both lightning strikes travel equal distances to reach the observer at the centre. Because the flashes arrive simultaneously, the observer concludes that the lightning strikes occurred at the same time.

But now consider an observer sitting inside the train at its centre. As the train moves forward, this observer is moving toward the light from the front strike and away from the light from the rear strike. As a result, the flash from the front reaches the observer earlier than the flash from the rear. From the perspective of the observer inside the train, the two lightning strikes did not occur simultaneously. One occurred earlier than the other.

Both observers are correct within their own frames of reference. This means that the idea of a universal "present moment" shared by the entire universe does not exist. Instead, the concept of "now" becomes relative, depending on the motion of the observer.

Next, we have another fascinating consequence of motion at extremely high speeds, The Relativistic Doppler Effect. To understand it, let us begin with something that you might have studied - the Doppler effect. This phenomenon was first described by the Austrian physicist Christian Doppler. It occurs whenever a wave source moves relative to an observer. A simple everyday example is the sound of an ambulance siren. As the ambulance approaches you, the pitch of the siren appears higher. As it moves away, the pitch suddenly becomes lower.

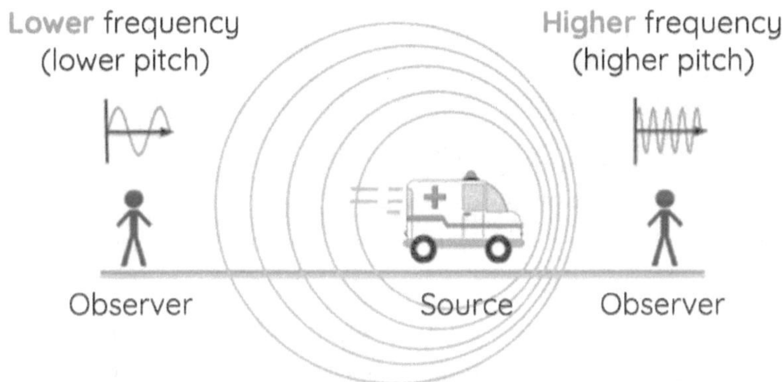

Figure 6.1: Visual Representation of the Doppler Effect using an Ambulance.
Source: scienceready.com The Doppler Effect

This happens because the motion of the source changes the spacing of the waves. When the source moves toward the observer, the waves become compressed, increasing their frequency. When the source moves away, the waves stretch apart, lowering their frequency.

The same principle applies to light waves. When a light source moves toward an observer, its wavelengths become compressed. This shifts the light toward the blue end of the spectrum, an effect known as blueshift. When the source moves away, the wavelengths stretch, and the light shifts toward the red end of the spectrum. This is known as redshift. At ordinary speeds, these changes are extremely small. But when velocities approach the speed of light, the effect becomes much more dramatic. In such situations, the shift must be described using the relativistic framework.

The relativistic Doppler effect takes into account not only the motion of the source but also the effects predicted by Special Relativity, including time dilation. The relationship describing the relativistic Doppler shift is often written as

$$f' = f\sqrt{1 + v/c \,/\, 1 - v/c}$$

Here,

f represents the frequency emitted by the source.

f' represents the frequency observed.

v is the relative velocity between the source and the observer.

c is the speed of light.

This equation shows how dramatically the observed frequency can change when velocities become comparable to the speed of light.

One of the most remarkable applications of the relativistic Doppler effect appears in astronomy. When astronomers observe distant galaxies, they often measure the spectrum of the light emitted by stars within those galaxies. When these spectral lines appear shifted toward longer wavelengths, astronomers know that the galaxy is moving away from us. When they appear shifted toward shorter wavelengths, the galaxy is moving toward us.

In the early twentieth century, the American astronomer Edwin Hubble carefully studied the light coming from distant galaxies. What he discovered changed our understanding of the cosmos. Nearly every distant galaxy exhibited a redshift. This meant that almost all galaxies were moving away from us. More astonishingly, the farther away a galaxy was, the faster it appeared to be moving away. This observation led to one of the most profound conclusions in cosmology: the universe itself is expanding. The light from distant galaxies is stretched as space itself expands, producing the redshift we observe. Without the Doppler effect, we would never have discovered this fundamental property of the universe.

Denouement VI

Covering all these postulates of Special Relativity, one fact remains: why do all these phenomena occur? And why is it the meticulous way? All these phenomena — length contraction, relativistic mass increase, time dilation, the relativity of simultaneity, and the relativistic Doppler effect — occur for a single fundamental reason: the speed of light is constant for every observer. To preserve this universal constant, the universe adjusts the measurements of space, time, energy, and simultaneity depending on the relative motion of observers. What all does the universe do? It shortens the distance light has to travel in a moving frame, ensuring the ratio of distance to time remains exactly 'c', leading to length contraction. It acts as an "inertial drag" that prevents any object from ever reaching or exceeding 'c', no matter how much energy is added, leading to relativistic mass increase. It shifts the "start" and "stop" times on an observer's clocks, perfectly cancelling out their own motion relative to the light beam, leading to relativity of simultaneity. It slows down the ticking of a moving clock so that even as distances shrink, the "seconds" stretch to keep the measured speed of light stable, leading to time dilation. And shifts the frequency and wavelength of light to compensate for the observer's motion while the actual propagation speed 'c' remains unchanged, leading to relativistic Doppler effect. This explains why it is called "The Meticulous Way"—because the universe upholds its fundamental laws with remarkable precision and consistency.

Quite a remarkable amount of knowledge has been processed in this chapter. We have seen how mass converts into energy, in other words, how mass is equivalent to energy and how motion reshapes space and time.

So, where do we head next? We head next to a chapter where we will explore the utmost limits of the universe. We will step toward those very boundaries. We will explore the fundamental barriers that nature imposes — the limits that prevent us from going further. What stops us? And what would happen if we dared to push beyond those limits?

7

The Barriers

"Even the light and gravity aren't instantaneous,
limited by the max speed barrier"

Every phenomenon in the universe operates within a maximum limit. No matter how powerful our technology becomes, no matter how far our imagination stretches, nature itself sets boundaries that cannot easily be crossed. But what exactly are these limits? Where do they come from? And what happens when we attempt to push against them? So here we are, and let us explore them.

Throughout the history of physics, scientists have repeatedly discovered that the universe does not allow quantities to grow without bound. There is always a ceiling — a point beyond which the familiar rules either break down or transform into something entirely different. These limits appear in many forms: the maximum speed at which information can travel, the smallest measurable units of space and time, the highest possible densities of matter, and the immense energies required to push nature beyond its comfort zone. These barriers are not merely inconveniences for scientists. They are fundamental features of reality. They define what can exist, what can happen, and even what can be known.

Consider, for instance, the limit we have already encountered multiple times in this book — the speed of light. This speed, represented by the constant *c*, is not simply the speed of a particular particle called a photon. Instead, it acts as a cosmic speed limit, shaping the structure of space and time themselves. As our understanding deepened through the twentieth century, scientists began uncovering other astonishing limits.

But why should such a limit exist at all? Why should the universe restrict motion in this way? What would happen if we tried to cross these barriers? There are many questions with very anonymous answers; let us step closer to the boundaries of the cosmos and discover where nature draws the line.

Causality

Every event in the universe appears to follow a simple and intuitive rule: causes come before effects. If a glass falls from a table, it breaks after it hits the ground. If a star explodes, the light from that explosion reaches distant observers only after the explosion has occurred. Our everyday experience constantly reinforces this idea. Yet this simple rule, which we rarely question, forms one of the deepest foundations of physics. This rule is known as the law of causality. So, what is the law of causality? Causality refers to the principle that an event (the effect) must always be preceded by something that produced it (the cause). In other words, information about a cause must travel through the universe before the effect can occur.

Our first barrier: How fast can that information travel? If causes must always precede effects, there must exist some maximum speed at which information, influence, or physical interaction can propagate through space. Without such a limit, the orderly chain of cause and effect could easily collapse. Modern physics tells us that this maximum speed is the speed of light in a vacuum, a central result of the Special Theory of Relativity.

The speed of light in vacuum is approximately: $c = 3 \times 10^8\ ms^{-1}$ This speed is denoted by the letter 'c.' While it is commonly referred to as the speed of light, its role in physics is much deeper. It is not merely the speed at which photons travel. Instead, it represents the maximum speed at which information, matter, or any causal influence can move through spacetime. In other words, nothing that carries information can travel faster than this speed. This fact transforms the speed of light into something far more fundamental than a property of a particular particle. It becomes a universal constant embedded in the structure of reality itself.

Why the letter 'c'? Historically, the symbol c originates from the Latin word 'celeritas', meaning swiftness. Early physicists adopted this notation when describing the propagation speed of electromagnetic waves. However, in modern physics, the meaning of 'c' has acquired a much deeper interpretation. Rather than simply representing the speed of light, the letter 'c' is often interpreted as the speed of causality, which is the maximum rate at which cause-and-effect relationships can propagate through the universe. Whenever one event influences another, the information about that influence must travel through spacetime. That transmission cannot occur faster than 'c'. Because of this role, 'c' acts as a boundary separating events that can influence each other from events that cannot. This idea becomes clearer when we consider how spacetime organises events, the Structure of Cause and Effect.

In the relativistic view of the universe, every event occurs at a specific location in space and time. Together, these coordinates form a point in spacetime. Spacetime can be imagined as a four-dimensional framework combining three dimensions of space and one dimension of time. As we had seen in Chapter 1, three spatial and one temporal dimension.

Within this framework, events are connected by what physicists call light cones. Imagine a flash of light occurring at a particular point in space and time. As the light spreads outward in every direction, it forms a spherical wavefront expanding at the speed c. If we represent this expansion in spacetime, the wavefront traces out a shape known as a light cone. The light cone divides spacetime into three distinct regions:

1. **The Past Light Cone**
   This region contains all events that could possibly influence the present event. Any signal travelling from these events toward the present must move at or below the speed of light.
2. **The Future Light Cone**
   This region contains all events that the present event can influence.
3. **The Elsewhere Region**
   Events outside both cones cannot influence the present event, nor can the present event influence them. Because doing so would require faster-than-light communication.

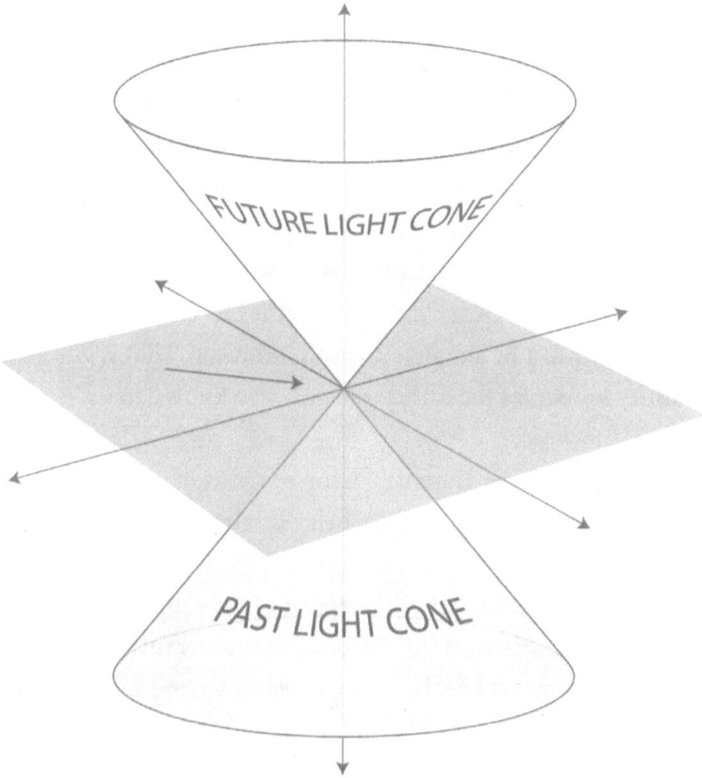

Figure 7.1: Light Cones
Source: Wikipedia.org

The structure of these light cones is what preserves causality. They ensure that causes remain confined to the past of their effects. But what would happen if something could travel faster than the speed of light? To answer this question, we must first understand how strongly nature enforces this cosmic speed limit. Let us examine these enforcers one by one.

The very first obstruction we face is the infinite energy demand. Imagine a spacecraft drifting in the silent vastness of space. Its engines ignite, pushing it faster and faster. At first, the process seems simple enough. Add more fuel, produce more thrust, and the spacecraft should continue to accelerate. In the classical mechanics of Isaac Newton, nothing prevents an object from accelerating indefinitely. If one continues applying force, velocity simply keeps increasing. But nature behaves very differently when speeds approach the speed of light. But we know closer the object comes to the speed of light, the more energy must be supplied to push it further. That is governed by a quantity known as the Lorentz factor, also known to us.

$$\gamma = 1 / \sqrt{1 - v^2/c^2}$$

At low velocities, the Lorentz factor remains extremely close to 1. Under these conditions, motion behaves almost exactly as classical physics predicts.

But as velocity approaches the speed of light, the denominator of this expression begins to shrink, causing the Lorentz factor to grow larger and larger. When v becomes extremely close to c, the Lorentz factor increases enormously. This growing factor appears directly in the expression for relativistic energy:

$$E = (\gamma - 1)mc^2$$

where m represents the object's rest mass. The equation reveals that energy grows without limit as velocity approaches the speed of light. To actually reach c, the Lorentz factor would have to become infinite. An infinite Lorentz factor would require infinite energy. Since no physical system can provide infinite energy, any object possessing rest mass can only approach the speed of light — never reach or exceed it.

Particle accelerators like the Large Hadron Collider provide direct confirmation of this behaviour. Particles are accelerated to speeds extremely close to c. Yet even with tremendous energy inputs, they never quite reach it. The closer they get, the harder it becomes to push them further. The speed of light, therefore, acts as an asymptotic boundary — a limit that can be approached but never crossed.

Energy alone is not the only barrier preventing faster-than-light motion. An even deeper reason lies in the preservation of causality. As mentioned, this simple rule governs every physical process in the universe. However, relativity reveals that the ordering of events in time can depend on the motion of the observer. Different observers moving relative to each other may disagree about which events occur simultaneously. Under normal circumstances, this disagreement does not create any paradoxes because no signal can travel faster than light. But if faster-than-light communication were possible, the situation would change dramatically. Relativity reveals something deeply surprising about time: the order of events can depend on the motion of the observer. Two observers moving relative to one another may not always agree about which events occur simultaneously. This strange feature is known as the relativity of simultaneity, a key consequence of the Special Theory of Relativity.

Under normal circumstances, this difference in perspective does not create any problems. The reason is simple: no signal can travel faster than the speed of light. Because of this limit, the sequence of cause and effect always remains consistent for every observer. But imagine a situation where faster-than-light (FTL) communication was possible. To see how dangerous this would be for causality, consider two observers:

Oggy and Jack - yes, from the cartoon Oggy and the Cockroaches.

Oggy is floating calmly in a spacecraft, essentially at rest Oggy is floating calmly in a spacecraft, essentially at rest with respect to a distant star system. Jack, however, is racing past Oggy in another spacecraft at an enormous speed, close to the speed of light. Because of their relative motion, Oggy and Jack measure space and time differently. Their clocks tick at different rates, and their definitions of "now" are not the same. Suppose Oggy observes a dangerous asteroid heading toward Jack's ship. Concerned, Oggy immediately sends Jack a warning message. If this message travels at or below the speed of light, everything behaves normally. In Oggy's frame of reference, the message is sent first and received later. In Jack's frame, the same order of events is preserved—the cause still precedes the effect. But now imagine the message could travel faster than light.

From Oggy's perspective, the situation still seems straightforward. He sends the warning message, and Jack receives it shortly afterwards. Cause is followed by effect. However, because Jack is moving at such high speed relative to Oggy, relativity reshapes Jack's perception of time. The ordering of distant events in his frame becomes different due to the relativity of simultaneity. In Jack's frame of reference, the faster-than-light signal might arrive before Oggy actually sends it. In other words, Jack could receive the warning message before the moment Oggy decides to transmit it. At first, this might seem like a harmless curiosity, but the consequences quickly become profound.

Imagine Jack receives the message early enough to respond. Realising that an asteroid threat exists, he immediately sends a reply message—also faster than light—back to Oggy. In his reply, Jack says:

"*Meow Meow:* Do not send the warning. There is no danger anymore."

Because this reply also travels faster than light, Oggy might receive Jack's message before he sends his original warning. Now Oggy faces a strange situation. He receives Jack's message telling him not to send the warning. If he obeys the message and never sends it, then Jack would never have received the original warning in the first place. But if Jack never received the warning, why would he send the reply telling Oggy not to send it?

The sequence of events becomes logically impossible. The effect—the warning reply—has eliminated its own cause.

This type of contradiction is known as a causality paradox, sometimes called a bootstrap paradox, where information or events appear without any clear origin in time. Such paradoxes would multiply endlessly if faster-than-light communication were allowed. Messages could be sent into the past, events could influence their own causes, and the orderly progression of time would collapse into logical chaos. Physics strongly suggests that the universe cannot permit such inconsistencies. Every experiment performed so far—from high-energy particle collisions to astronomical observations—confirms that no information travels faster than the speed of light.

The cosmic speed limit, therefore, acts as a protective boundary.

Now that we have understood a mind-bending topic in a very easy way, let us look at how relativity enforces the cosmic speed limit through another remarkable mechanism: the distortion of space and time themselves. We know that as velocity increases, time dilation and length contraction begin to appear. Time dilation, as discussed in Chapter 2, means that moving clocks run more slowly relative to stationary observers. The relationship is given by

$$\Delta t = \gamma \Delta \tau$$

where

$\Delta\tau$ represents the proper time measured by the moving observer.

As velocity approaches the speed of light, the Lorentz factor grows extremely large, causing time to slow dramatically.

Length contraction produces the complementary effect:

$$L = L_0/\gamma$$

where L_0 represents the proper length.

Objects moving at relativistic speeds appear shortened along the direction of motion. These effects have been experimentally confirmed through numerous experiments involving high-speed particles and precise atomic clocks. And we know these very well from the chapters read before. Yet as velocity approaches c, these distortions become extreme.

Time dilation grows enormous.

Lengths shrink dramatically.

And the mathematics itself begins to resist further acceleration.

If an object were somehow pushed beyond the speed of light, the Lorentz factor would become imaginary ($\sqrt{-x} = i$) — leading to physically meaningless results. In other words, spacetime itself prevents such motion.

Okay, we understand that light's speed barrier imposes restrictions when we try to approach its speed, but if we don't, does it still affect anything?

Related to this, let me tell you an interesting fact. The cosmic speed limit imposed by relativity does not influence only spacecraft or particles in accelerators. Its consequences reach even into the quiet structure of atoms themselves. In fact, the very existence of the chemical elements may ultimately be limited by the same barrier we have been discussing.

Consider the familiar Periodic Table. Today, it contains 118 confirmed elements, the heaviest being Oganesson. For more than a century, scientists have extended the table step by step, synthesising heavier and heavier nuclei in powerful laboratories. At first glance, it might seem that the table could continue indefinitely if we simply kept adding more protons to the nucleus. However, the laws of physics hint that this growth cannot continue forever. Current theoretical studies estimate that the ultimate maximum number of elements may lie somewhere between 137 and 173.

The Feynman Point

The celebrated physicist Richard Feynman pointed out an intriguing consequence of this idea. In a simplified atomic model originally proposed by Niels Bohr, the velocity of an electron in the innermost orbit is proportional to the atomic number 'Z' multiplied by a fundamental constant known as the fine structure constant.

This constant, usually written as α, has a value very close to

$$\alpha \approx 1/137$$

Within the Bohr model, the approximate velocity of the innermost electron can be expressed as

$$v \approx Z\alpha c$$

where

Z is the atomic number,

α is the fine structure constant, and

c is the speed of light.

For ordinary atoms, this velocity remains well below the cosmic speed limit. But something fascinating occurs as the Z grows larger.

When Z = 137, the equation predicts

$$v \approx c$$

In other words, the electron in the innermost orbital would need to travel exactly at the speed of light to maintain its orbit. But relativity forbids any particle with mass from reaching that speed. If we attempted to go beyond Z = 137, the mathematics of the Bohr model would require the electron to travel faster than light—something the laws of physics simply do not allow.

Because of this reasoning, physicists sometimes refer to Z = 137 as the Feynman point, and the hypothetical element at this position has occasionally been nicknamed "Feynmanium." Of course, the Bohr model is only a simplified description of atomic structure. Modern quantum mechanics provides a far more accurate framework. Yet the appearance of this limit remains a powerful hint that the periodic table may not extend indefinitely.

The Dirac Prediction

When the more complete relativistic quantum theory developed by Paul Dirac is applied to heavy atoms, the story becomes slightly more subtle. The Dirac Equation incorporates both quantum mechanics and relativity, providing a more accurate description of electron behaviour in extremely strong nuclear fields.

Dirac's analysis suggests that atoms might remain mathematically consistent even beyond Z = 137. However, the equations begin to behave strangely as the atomic number continues to grow. When the nuclear charge approaches roughly Z = 173, the electric field surrounding the nucleus becomes so intense that the vacuum itself becomes unstable. At this point, the theory predicts that the energy levels of electrons plunge into the negative energy continuum, allowing the vacuum to spontaneously produce electron–positron pairs. Ya, it's the same matter and anti-matter pair which we discussed in Chapter 5.

In simpler terms, the electromagnetic field of such an enormous nucleus would become strong enough to tear particles directly out of empty space. This phenomenon indicates that atoms with Z greater than about 173 cannot exist as ordinary chemical elements. The structure of the vacuum itself would prevent them from forming stable atomic systems.

Beyond Causality

Throughout our way in this chapter, we have seen that the universe guards the speed of light as a strict boundary. Energy demands become infinite, causality begins to fracture, and spacetime itself resists any attempt to cross the limit. But suppose, purely as a thought experiment, completely hypothetical, that a traveller somehow managed to exceed this boundary. What would such a journey look like? What would the traveller actually experience? The answer is far stranger than simple, faster motion. If the cosmic speed limit were somehow violated, the traveller would encounter a series of bizarre and dramatic consequences—some subtle, some catastrophic. Let us explore the strange price one would pay.

One of the most fascinating consequences of extremely high speeds is the possibility of viewing the past. It is important to clarify this carefully. This does not mean travelling back in time or altering historical events.

So don't think about time travel and creating a 'Grandfather paradox'. Instead, it simply means observing light that left Earth long ago, which is still travelling through space. To understand how this works, we must remember a simple fact about light. Light travels at a finite speed. Whenever an event occurs on Earth, the light carrying information about that event spreads outward into space in all directions at the speed c. This means that the universe is filled with expanding shells of information from every historical event. If we move far enough away from Earth, we can intercept light that left the planet years—or even decades—ago. In other words, we can see the past. Astronomers already do this on cosmic scales. When we observe distant galaxies, we are seeing them not as they exist today, but as they existed millions or billions of years ago. In simple terms, when you look at a star in the night sky, that star might have died, but it is still visible because the light coming from that star is taking a finite time to reach the Earth. The same principle applies to Earth itself.

Let us consider a striking historical event: the Atomic bombing of Hiroshima during World War II. The explosion occurred on 6 August 1945. The light from that event immediately began travelling outward into space at the speed of light. Even today, that light continues its journey. Suppose a spacecraft left Earth today and somehow travelled faster than light. If it moved outward far enough and then looked back toward Earth, it could intercept the light that departed decades ago. How far away would we need to be? Let us estimate. From 1945 to 2026 is approximately 81 years.

Since light travels one light-year per year, the light from the Hiroshima explosion is currently about 81 light-years away from Earth. Well, that's a lot, not if you're travelling with an obvious miraculous speed. For your reference, the closest star to our Solar System, Proxima Centauri, is approximately 4.25 light-years away from Earth. And the most distant man-made object, Voyager 1, is approximately just 0.0027 light-years away. Coming back to the topic, this means that if an observer could instantly travel 81 light-years away and look back with a powerful telescope, they would see Earth exactly as it appeared in 1945. They would witness the moment of the bombing unfolding in real time. But they would only be watching history, not influencing it. The event already occurred; the observer is merely catching up to the light that recorded it.

So, we know what the observer can see, but now let us imagine what the traveller would actually see while moving faster than light. Surprisingly, the answer is not simply that everything would blur past. In fact, vision itself would begin to break down. Three dramatic visual effects would appear.

First, consider the region behind the traveller. Normally, we see objects behind us because light from those objects eventually reaches our eyes. But if the traveller were moving faster than the light coming from behind, those photons would never catch up. The traveller would literally outrun the light. As a result, everything behind the spacecraft would disappear into darkness. The region of space behind the traveller would become a perfect visual void. No stars, no galaxies, nothing. It would be as though the universe had simply vanished in that direction.

Second, the situation ahead of the traveller would be the opposite—and far more dangerous. Light coming from the direction of motion would become dramatically compressed due to the relativistic Doppler effect. The Doppler effect occurs whenever a source and observer move relative to each other. For example, the pitch of a passing siren increases as the source approaches and decreases as it moves away, like the ambulance analogy we had seen. Light behaves in a similar way, except that the shift occurs in frequency and colour. If an observer moves toward a light source, the wavelength shortens, and the light shifts toward the blue end of the spectrum. This phenomenon is known as blue-shift. We know this, but at extreme speeds, this effect becomes astonishingly powerful.

Visible starlight would be compressed so intensely that its frequency would increase far beyond the visible spectrum. Ordinary light would shift into ultraviolet radiation, X-rays and gamma rays. Instead of seeing stars as gentle points of light, the traveller would face a wall of extremely high-energy radiation. Gamma rays are powerful enough to damage living tissue and disrupt molecular structures. How powerful are they? Gamma rays can penetrate even a 30 cm or more dense iron wall. A spacecraft moving at such speeds would be bombarded by intense radiation capable of destroying electronics and harming biological organisms.

A third visual effect would also occur, known as relativistic aberration. At very high speeds, light from all directions becomes distorted by the geometry of spacetime. Instead of arriving from its usual directions, the light rays bend toward the direction of motion. As velocity increases, the entire visible sky appears to collapse into a narrow cone directly in front of the traveller. Stars that were originally above, below, or even slightly behind begin to appear concentrated in the forward direction. Eventually, the entire universe would seem compressed into a single brilliant point ahead. The traveller's field of view would shrink dramatically, leaving only a tiny, intensely bright region in front. Everything else would fade into darkness. The universe itself would appear to focus into a single dazzling spot.

Now that we know how the visuals would look, there is another, even stranger possibility. When particles travel faster than light in a material medium, they produce a phenomenon known as Cherenkov

radiation.
Although nothing can exceed the speed of light in a vacuum, light actually travels more slowly through materials such as water or glass. In such environments, certain high-energy particles can exceed the local speed of light. When this happens, the particle produces a distinctive blue glow, similar to a sonic boom but with light. It is commonly observed in nuclear reactors, where high-energy electrons move through water faster than light travels in that medium. When this "over-speeding" happens, it creates an electromagnetic shockwave known as Cherenkov Radiation. This is why nuclear reactor cores glow with a ghostly, brilliant blue light.

Imagine the implications if an object could exceed the speed of light in the vacuum itself. Physicists suspect that something similar, but far more extreme, might occur. The vacuum of space is not truly empty. According to quantum physics, it is filled with fluctuating energy fields. If an object were somehow forced to move faster than 'c', it might interact violently with these fields. Instead of emitting Cherenkov radiation into water, it could begin emitting radiation directly into the vacuum of space. The energy loss could become enormous.

In fact, theories suggest the energy leakage might become effectively infinite, causing the object to shed energy uncontrollably. In extreme scenarios, such a disturbance might even trigger a vacuum phase transition, a process in which the quantum state of space itself changes. Such a transition could potentially release tremendous energy and destabilise the surrounding region of spacetime.

Lastly, we face the very familiar consequence, the Lorentz factor. When $v<c$, the expression under the square root remains positive. But if v exceeds c, something strange happens. The quantity inside the square root becomes negative. The result is no longer a real number; it becomes an imaginary number. In physics, imaginary numbers often indicate that a theory has entered a regime that ordinary physical

objects cannot occupy. If an object with rest mass were forced beyond the speed of light, its relativistic mass would become mathematically imaginary.

Particles that would naturally exist in such a regime are sometimes called tachyons, hypothetical faster-than-light particles that have never been observed experimentally. The appearance of imaginary mass suggests that objects travelling faster than light would behave in fundamentally different ways from ordinary matter. Their energy would decrease as their speed increased, and slowing them down to the speed of light would require infinite energy. This strange behaviour further reinforces the idea that crossing the speed of light is not simply difficult—it is fundamentally incompatible with the physics of ordinary matter. And finally, we have covered one of the biggest barriers - the speed barrier of our universe.

Ember and Ice

At the end of our previous discussion, we encountered one of the universe's most fundamental restrictions: the speed limit of light. But speed is not the only boundary that nature imposes. Just as the universe places a limit on how fast something can move, it also imposes limits on how hot or how cold something can become.

Temperature, at its deepest level, is simply another expression of motion. When particles move rapidly, collide violently, and vibrate intensely, we describe the system as hot. When motion slows down, collisions weaken, and vibrations fade, we describe the system as cold. Thus, temperature is, in a sense, a measure of the average energy of motion within matter.

And if motion itself has limits—as we saw with the speed of light—it is natural to ask whether temperature must also have limits. Indeed, it does. The universe operates between two astonishing extremes. At one end lies a state where motion is reduced to its smallest possible level, where matter approaches perfect stillness. At the other end lies a state where energy becomes so intense that the known laws of physics dissolve into a chaotic mathematical storm. Let us begin with the coldest possible state of reality.

Imagine cooling a substance continuously. At room temperature, atoms inside the material vibrate rapidly, bouncing around like restless dancers. As the temperature drops, these motions begin to slow. The atoms vibrate less violently, their movements becoming more subdued. Lower the temperature further, and the motion slows even more. Eventually, we approach a remarkable theoretical limit known as 'absolute zero'. Something you must have studied in your Physics textbooks. Absolute zero corresponds to:

0 Kelvin = -273.15 °C = -459.67 °F

The fascinating part now starts, at this temperature, all classical thermal motion would theoretically stop. Atoms would no longer jiggle randomly due to heat. Molecules would cease their thermal dance. The entire system would settle into the lowest possible energy state. It is, in a sense, the limit of stillness. But nature has placed an important restriction here as well. According to the Third Law of Thermodynamics, it is impossible to reach absolute zero through any finite series of physical processes. We can approach it arbitrarily closely, but we can never actually arrive there. This means that no laboratory, no experiment, and no cosmic process can perfectly remove all thermal energy from a system. There will always remain a tiny residual motion.

Even the most advanced experiments on Earth—using lasers, magnetic traps, and sophisticated cooling techniques—have only managed to reach temperatures a few billionths of a Kelvin above absolute zero. In these extreme environments, matter begins to behave in astonishing ways. Atoms lose their individual identities and begin acting collectively, forming exotic states such as Bose–Einstein condensates (the fifth state of matter), where particles behave as though they are part of a single giant quantum object. Yet even here, absolute stillness remains unreachable. Here's a note on what BEC is and one of its most prominent examples.

When certain atoms are cooled to temperatures extremely close to absolute zero, their individual quantum waves begin to overlap. Instead of behaving like separate particles, they merge into a single collective quantum state. This phenomenon was predicted by Satyendra Nath Bose and Albert Einstein and is therefore called a Bose–Einstein condensate. In this state, thousands or even millions of atoms behave as though they are one giant "super-atom," moving in perfect quantum synchrony. It is one of the clearest demonstrations that matter at extremely low temperatures is governed not by classical physics, but by quantum mechanics. A striking example occurs with helium cooled to just a few degrees above absolute zero. At these temperatures, helium can enter a phase known as Superfluidity. In this state—often observed in Helium-4—the liquid loses all internal friction and flows with zero viscosity. It can creep up the walls of a container, pass through microscopic pores that ordinary liquids cannot, and continue circulating indefinitely without losing energy. Like BECs, this behaviour arises because the atoms begin acting collectively as a quantum fluid rather than as independent particles.

When we try to understand why perfect stillness is forbidden, we again encounter one of the most profound principles in modern physics:

the Heisenberg Uncertainty Principle. Let's try to understand it now. Imagine what would happen if atoms truly reached absolute zero. Their motion would completely stop. Their momentum would become exactly zero. If that happened, we would know both their position and momentum with perfect precision. But the uncertainty principle forbids such perfection. To preserve the rules of quantum mechanics, particles must always retain a tiny amount of residual energy. This unavoidable energy is known as zero-point energy. They cannot remain perfectly still. But, if a particle reached 0 K and stopped entirely, we would know its momentum is exactly zero, which would cause its position to become "infinite." Therefore, the universe enforces a "Zero-Point Energy" that keeps things jittering just enough to exist.

Now, let us travel to the opposite end of the cosmic thermometer. Just as the speed of light creates a boundary for velocity, the universe appears to possess a maximum possible temperature. Beyond this point, the very framework of physics begins to fail. This ultimate boundary is known as the 'Planck Temperature', denoted by T_P. It is defined by the equation:

$T_P = \sqrt{} \hbar c^5 / G k^2$

Where:

$\hbar$ = reduced Planck constant

c = speed of light

G = gravitational constant

k = Boltzmann constant

When these fundamental constants combine, they produce an astonishing number:

$T_P \approx 1.417 \times 10^{32}$

This is one hundred nonillion Kelvin.

To appreciate how extreme this is, consider the following comparisons:

The surface of the Sun is about 5800 K.

A supernova explosion may reach billions of Kelvin.

Particle accelerators briefly create temperatures of trillions of Kelvin.

The hottest place in the universe (the core of the quasar 3C 273) is

10

trillion Kelvin.

Yet even these colossal values are unimaginably small compared to the Planck temperature. The Planck temperature represents what might be called the 'absolute hot' of the universe. At this point, energy densities become so enormous that the fundamental forces of nature begin to behave in ways we cannot currently describe.

Physicists believe that the universe briefly reached this temperature during the earliest instant of the Big Bang. This period is known as the 'Planck Epoch', lasting roughly:

10^{-43} seconds

after the beginning of the universe. At that moment, the entire observable universe was compressed into an incredibly tiny region filled with unimaginable energy. Temperatures soared to the Planck scale. Under such conditions, the distinction between the fundamental forces likely vanished. Gravity, electromagnetism, and the nuclear forces may have merged into a single unified interaction. The universe existed as an almost incomprehensible cosmic soup of energy and spacetime fluctuations.

I think we should try breaking these limits too, as we did in the case of light. But to understand what happens beyond the Planck temperature, physicists believe we need an entirely new framework. A theory that successfully unifies gravity with quantum mechanics. Such a theory is often called a Theory of Everything. Candidates include ideas such as string theory and loop quantum gravity, but none have yet been experimentally confirmed. So, for now, we can count on some pre-existing data and a pinch of hypothesis.

Let's attempt to cool a particle below 0 Kelvin. At first glance, this might sound like simply making something "colder than the coldest." However, we know that temperature is not just a number on a scale; it is deeply connected to the kinetic energy of particles. Temperature is proportional to the average kinetic energy of the particles in a system. As the temperature decreases, this average kinetic energy decreases as well. At absolute zero, the system has reached its lowest possible energy state. Going further below, the mathematics of ordinary thermodynamics would suggest something impossible. The kinetic energy would become negative. But kinetic energy cannot physically be negative. A particle cannot possess "less than zero motion." Motion can decrease, but it cannot become "less than not moving." Thus, attempting to cross absolute zero in ordinary physics results in what can be described as a mathematical error. It would imply that particles are moving less than not at all, which is meaningless within standard physics. This is one of the reasons absolute zero acts as a hard boundary in thermodynamics. Thus, when viewed through the lens of standard thermodynamics, going below absolute zero is not simply difficult—it is physically undefined.

Travelling to the opposite extreme, we find the Planck temperature, and our goal is to surpass it. So, let's add more energy to it. At first glance, one might assume the temperature would simply increase further. However, something remarkable occurs when we examine the situation more closely. In quantum physics, particles behave not only like particles but also like waves. The wavelength associated with a particle is described by the relation

$$\lambda = h/p$$

where

h is Planck's constant

p is the particle's momentum

As the energy of a particle increases, its wavelength becomes smaller and smaller. At extremely high energies—approaching the Planck temperature—the wavelength of particles shrinks toward the Planck length. We know that the Planck length represents the smallest meaningful scale in known physics. Now imagine adding more energy. The particle's wavelength would attempt to shrink below the Planck length. But here gravity suddenly becomes crucial. How?

We know energy and mass are equivalent, as described by our

mass-energy equivalence formula: $E = mc^2$. This means extremely energetic particles effectively possess enormous mass-energy density. When energy becomes concentrated within a sufficiently small region, gravity can become so strong that it forms a black hole. The critical size at which this occurs is given by the Schwarzschild radius, which defines the boundary of a black hole. If the Schwarzschild radius of the particle becomes larger than its wavelength, something extraordinary happens. The particle collapses under its own gravitational influence. Instead of becoming hotter or more energetic, it transforms into a tiny black hole. In other words, the universe prevents us from exceeding the Planck temperature by converting excess energy into gravity. The energy we try to add is effectively swallowed.

There is an interesting fact about such kind of black holes. A black hole formed purely from concentrated energy is sometimes referred to as a Kugelblitz. The term comes from German and literally means "ball lightning." Unlike ordinary black holes, which form from collapsing stars, a Kugelblitz would form entirely from intense radiation or energy. If enough energy were compressed into a sufficiently small region of space, gravity would force that energy to collapse into a black hole. Thus, attempting to heat a particle beyond the Planck temperature might not create a hotter particle at all. Instead, it could create a microscopic black hole made entirely of energy. In this sense, gravity acts as a cosmic safety mechanism, preventing temperatures from rising beyond a certain point.

Zenith of Radiance

So, in the previous segment, we explored the thermal extremes of the universe. On one end lies the stillness of absolute zero, where on the other end stands the inferno of the Planck temperature. And now, we come to brightness. If heat has its limits, so must brightness.

Stars shine, galaxies glow, and quasars blaze across the cosmos with astonishing luminosity. Yet even here, the universe does not allow unlimited freedom. Just as there exists a maximum speed and a maximum temperature, there is also a maximum brightness. This ultimate boundary is known as the
'Planck luminosity'.

The Planck luminosity represents the highest possible rate at which energy can be emitted according to the fundamental constants of nature.

It is given by the remarkably simple relation:

$L_P = c^5 / G$

Where:

L_P = Planck luminosity

c = speed of light

G = gravitational constant

When calculated, this yields

$L_P \approx 3.6 \times 10^{52}$ watts

Again, to appreciate the scale of this number, let us compare it with familiar cosmic sources of light.

The Sun emits roughly 4×10^{26} watts.

A large galaxy might emit around 10^{36} watts.

A powerful supernova explosion can briefly reach 10^{44} watts.

Yet the Planck luminosity surpasses even these cataclysmic events by many orders of magnitude. It represents an almost unimaginable level of energy release. At this brightness, energy would be pouring into space so rapidly that gravity itself could not remain passive.

A simple question: Why does such a limit exist? Logically, brightness should have no limit. But the thing is, the Planck luminosity arises from the same constants that define gravity and relativity. It represents the scale at which energy output becomes so enormous that spacetime itself must respond. If energy is emitted at a tremendous rate, the energy itself contributes to gravitational effects. At sufficiently extreme luminosities, the flow of energy through space would carry enormous gravitational influence. Instead of merely producing light, such an event would begin warping spacetime violently. To understand this better, try to imagine what would happen if we tried to cross this limit, with our "pinch of hypothesis".

Suppose, purely hypothetically, that some event attempted to exceed the Planck luminosity. The consequences would likely be extraordinary. Since luminosity represents energy released per unit time, exceeding this boundary would mean that energy is being pumped into spacetime at a rate too great for spacetime to remain stable. Something similar to the case of temperature would happen. One possibility is that the immense concentration of energy would immediately create intense gravitational fields, potentially forming microscopic black holes. This is what is meant by the wrapping of spacetime. Another possibility is that inside the dungeons of quantum mechanics, spacetime itself would undergo extreme fluctuations. At Planck scales, many theories suggest that space behaves like a turbulent structure known as quantum foam, where the geometry of spacetime constantly fluctuates. Beyond the Planck luminosity, this foam-like structure might erupt into violent distortions, making the concept of a smooth spacetime meaningless.

Denouement VII

Throughout this chapter, we have journeyed along the edges of the universe's limits—places where physics stops behaving like the comfortable rules we experience in everyday life and begins revealing its deeper architecture.

We began with the speed limit of the cosmos, the speed of light. This constant, represented by 'c', is not merely the speed at which light travels; it is the speed at which cause and effect themselves propagate through the universe. From there, we explored the strange consequences of approaching and hypothetically crossing this boundary. Then we turned toward the thermal limits of nature. At the coldest extreme lies absolute zero; at the opposite extreme stands the Planck temperature, the hottest conceivable state of matter. Finally, we lifted our eyes toward another boundary—the maximum brilliance the universe can produce, through the concept of Planck luminosity. Brightness being the last in the list but not the least, we finally exhaust the 'limit-list' of our universe.

But physics, like exploration, does not always proceed in neat and orderly chapters.

Some ideas emerge unexpectedly, concepts too intriguing to ignore, yet too unconventional or brief to stand as chapters of their own. Some appear as speculative theories, others as curious stories from the history of physics, and a few as thought experiments that stretch the imagination beyond familiar territory.

The next chapter is not a conventional one. It is a collection of such ideas, small windows into different corners of physics. Some will introduce speculative perspectives, some will recount remarkable insights from physicists, and others will present fascinating puzzles that challenge our intuition. In essence, it is a gallery of concepts that were too large to ignore and too small to become chapters themselves...

8

The Residual Glow

"The brightest fires are often fueled by the things we left behind."
brightest fires are often fueled by the things we left behind.

We, in our way, have discussed a variety of vast and intriguing ideas in physics—concepts that attempt to explain the universe with increasing clarity. From dimensional complexity, behaviour of time and space to the limits of the universe, each chapter has taken us deeper into the architecture of nature. Yet, along this journey, a few fascinating ideas remained behind. Not because they were unimportant, quite the opposite. Some of them are remarkably profound. But they exist in an unusual space: too large to be ignored, yet too small to form entire chapters of their own.

Some of these are stories, moments in scientific history born from unexpected observations, or even tragic accidents that revealed new truths about nature. Others are theories and thought experiments, ideas so imaginative that they initially seem almost unreal, yet so logically sound that they demand serious attention. So why leave these ideas unexplored?

Let us begin with one of the most famous thought experiments in the history of physics—an idea simple enough to imagine, yet powerful enough to reshape our understanding of reality.

Apple In A Box

Imagine a sealed wooden box, no larger than a loaf of bread, drifting silently in the vast emptiness of space. Inside the box lies a single apple; fresh, red, and firm, carrying the quiet fragrance of ripened fruit. There is nothing else inside: no air entering or leaving, no light passing through the walls, and no external influence disturbing what lies within. The box is perfectly sealed from the universe.

At the moment it is placed inside, the apple is perfectly intact. Its cells are organised into neat biological structures. Chemically, sugars, proteins, and water molecules are arranged in precise patterns that nature carefully assembled during the fruit's growth on a tree. In other words, or physically, the apple exists in a state of high order. But now imagine leaving this box alone.

Hours pass.

Days pass.

Weeks pass.

Gradually, the apple begins to change. Now, obviously, you know what kind of change will occur. The once-tight skin will soften. Internal chemical reactions will begin to break complex molecules. Microorganisms that may have been present on the surface begin their quiet work of decomposition. The fruit bruises, collapses and begins to rot. Over time, mould will spread through its flesh. The structured tissues will dissolve into a mixture of liquids and gases. Finally, the elegant architecture of the Apple will disappear completely. And eventually, what remains is no longer recognisable as fruit at all, just a chaotic mixture of decomposed organic matter. In everyday life, this process seems perfectly natural. We expect things to decay. We expect order to dissolve into disorder. This is precisely the phenomenon we encountered earlier in the book while discussing entropy in Chapter 2. It describes the tendency of systems to move from organised states toward more disordered ones, summarised by the Second Law of Thermodynamics, which states that in an isolated system, entropy tends to increase with time.

Let me tell you the reason why it is a thought experiment, which starts after this question: What if the story does not end with the rotten apple? Didn't get that? What if the box is left sealed for an unimaginably long time?

The molecules inside the box will rearrange themselves again, until the apple slowly reassembles itself back into its original fresh form. This is what would happen, or happens, who knows? At first, this sounds absurd. It feels like something that belongs in fantasy rather than physics. But surprisingly, the mathematics of statistical mechanics suggests that such a scenario is not strictly impossible. This strange possibility forms the heart of what we may call the 'Apple in a Box Theory'.

Before diving deeper, we must revisit the concept of entropy briefly. In Chapter 2, we saw that entropy is closely related to the number of microscopic configurations a system can have. In statistical mechanics, entropy is expressed through the famous relation proposed by Ludwig Boltzmann:

$$S = k \ln \square W$$

Where:

S = entropy of the system

k = Boltzmann's constant

$\ln$ = natural logarithm

W = number of microscopic configurations (microstates)

An idea, simple, yet powerful. A highly ordered system, like a fresh apple, has relatively few possible arrangements of atoms that correspond to that state. The molecules must sit in very specific positions to maintain the structure of the fruit. A decomposed apple, however, can exist in an enormous number of microscopic arrangements. The molecules can scatter and mix in countless ways while still appearing as a shapeless mush. Because there are vastly more disordered configurations than ordered ones, systems naturally evolve toward disorder.

In the late nineteenth century, the mathematician Henri Poincaré proved a remarkable theorem now known as the Poincaré Recurrence Theorem. What does it state? The theorem states that certain isolated systems, if left alone long enough, will eventually return to a state extremely close to their original configuration. The key conditions are:

The system must be finite.

The system must be isolated.

Energy must remain conserved.

Under these conditions, the system cannot wander through an infinite number of possible states. Eventually, its microscopic configuration must repeat. This idea forms the mathematical backbone of the Apple in a Box thought experiment. Inside the sealed box, the apple's atoms cannot escape. They remain confined within a finite volume. Although the number of possible arrangements of atoms is astronomically large, it is still finite. Given an infinite amount of time, the system must eventually revisit earlier configurations—including the one corresponding to a perfectly fresh apple. Our thought experiment rests upon three fundamental assumptions.

(i) Finite Phase Space

The apple and everything inside the box constitute a closed system. All the atoms—carbon, hydrogen, oxygen, nitrogen, and others—are confined within the boundaries of the box. In physics, we often describe the complete state of a system using a concept called phase space. Phase space represents all possible combinations of positions and momenta of the particles in a system.

Although the number of possible states for the apple is unimaginably large, quantum mechanics tells us that energy levels are quantised. This means the number of possible configurations is extremely large but ultimately finite. Like its said, the stars in the sky are uncountable, yet finite. The apple's atoms can only arrange themselves in a limited set of possible ways.

(ii) Microscopic Reversibility

At the microscopic level, the laws governing particle motion are reversible. If molecules collide, their interactions follow deterministic equations that work both forward and backwards in time. This idea is closely connected to the foundations of statistical mechanics developed by Ludwig Boltzmann. Boltzmann recognised that entropy emerges not from strict physical necessity but from statistical probability. Particles simply tend to explore the most numerous configurations available. For example, if a cup fell out of your hand and broke into pieces, you can say,

"Its molecules wanted to explore more configurations", until your mom is standing there.

(iii) Infinite Time

The final assumption is perhaps the most dramatic. We imagine that the box exists for an unlimited amount of time. If the system continues evolving indefinitely, it will eventually explore every accessible configuration within its phase space. And if that happens, the configuration corresponding to a perfectly fresh apple must eventually occur again.

Although the recurrence is mathematically inevitable, the timescale required is almost incomprehensibly large. For a system as complex as an apple, physicists estimate the recurrence time to be on the order of

$10^{10\text{\^{}}26}$years

It represents 1 followed by a 100 septillion zeroes. This number is so vast that it cannot be written down in decimal form. Even if you could write one zero on every single atom in the observable universe, you would run out of atoms long before finishing the zeros for this number.

For comparison:

The age of the universe is about 10^{10} years.

Proton decay, if it occurs, may happen after around 10^{34} years.

Yet the recurrence time for the apple is vastly larger than even these cosmic timescales. In practical terms, the apple will never reform. But mathematically, it eventually must. Why exactly such an absurdly enormous number? This number is so vast that it cannot be written down in decimal form. Even if you could write one zero on every single atom in the observable universe, you would run out of atoms long before finishing the zeros for this number. At this point, you might wonder: Why exactly such an absurdly enormous number? The answer lies in the staggering number of ways the atoms of the apple can arrange themselves.

A single apple contains roughly 10^{25} atoms, and each atom can occupy countless possible positions and energy states. When all these possibilities are combined, the number of microscopic configurations becomes astronomically large—far exceeding numbers normally encountered in physics. The system must wander through this immense landscape of states before returning to one that closely resembles the original apple. The recurrence time, therefore, grows to an almost inconceivable scale, roughly estimated as 10^{26} years.

We're almost done, but a mischievous question arises: What if, somehow, we open up the box in between? And it is quite interesting. So, opening the box allows entropy to increase even further. Instead of a closed system with fixed energy, the apple becomes part of a much larger thermodynamic system, the surrounding universe.

The probability of spontaneous reassembly becomes even more astronomically small. To understand this, imagine shuffling a deck of cards. The chance that a shuffled deck randomly returns to perfect original order is already extremely tiny. Now imagine someone suddenly adds another thousand cards into the deck and continues shuffling. The chance of returning to the original order becomes effectively zero. Opening the box is similar- it dramatically enlarges the system.

This thought experiment ends here. So what to conclude from it? We can conclude something very profound. The Second Law of Thermodynamics is statistical, not absolute. Entropy almost always increases because disorder is overwhelmingly more probable. But at the deepest level of physics, the laws themselves do not forbid the reverse.

Radiance of the Restless

Now, we step into one of the most extraordinary stories ever recorded in the history of science. But before that, we must first understand the phenomenon that lies at its centre - Radiation.

A word heard almost every day, encountered in plenty of subjects and mentioned quite a lot of times in this book. So what exactly is it? It is a word that often carries a sense of danger. In everyday language, people associate radiation with nuclear disasters, medical treatments, or invisible hazards. Yet radiation is not merely something that appears in extreme environments. In fact, radiation is one of the most fundamental and omnipresent phenomena in the universe.

At its simplest level, radiation is the transfer of energy through space or through a medium in the form of waves or particles. This energy may travel across the vacuum of space or pass through materials. The defining feature is that radiation moves energy from one place to another without requiring direct physical contact between objects. When the Sun warms the Earth, the energy does not travel through air currents or direct touch. Instead, it arrives in the form of radiation. Similarly, when a glowing piece of metal emits heat and light, it is radiating energy outward into its surroundings. Thus, whenever energy spreads outward from a source through waves or particles, we describe that phenomenon as radiation. But this definition opens an even deeper question. What exactly carries this energy? Radiation can exist in two fundamental forms.

1. Electromagnetic radiation
2. Particle radiation

Both carry energy, yet they behave very differently.

The most familiar form of radiation is electromagnetic radiation. This type of radiation consists of oscillating electric and magnetic fields travelling through space at the speed of light. These waves require no material medium to propagate, which is why sunlight can travel across the vast vacuum separating Earth from the Sun. Light is an electromagnetic wave, but visible light represents only a small fraction of a much larger family known as the electromagnetic spectrum. Yes, it is the same familiar spectrum that you must know, which we studied in the Doppler effect.

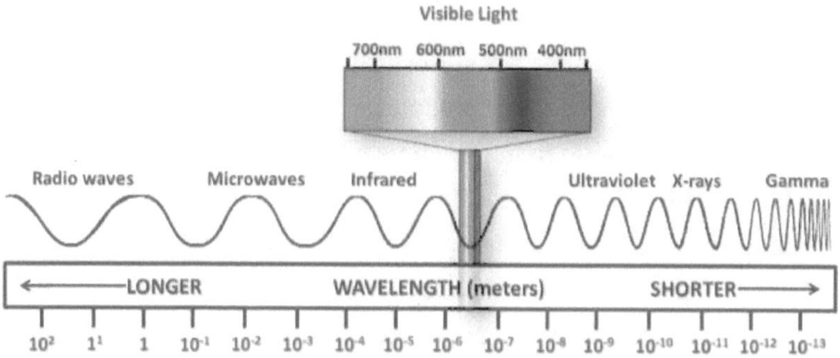

Figure 8.1: The Electromagnetic Spectrum
Source: NASA Climate Science Investigation - Electromagnetic Radiation

The spectrum includes radiations in many forms depending on their wavelength and frequency. It includes:

- **Radio Waves** - They possess the longest wavelengths and the lowest frequencies in the electromagnetic spectrum, widely used in communication systems such as broadcasting and satellite transmissions.
- **Microwaves** - They are commonly associated with microwave ovens, where they interact with water molecules in food, causing them to vibrate and produce heat. Yet microwaves are not merely a kitchen convenience. The faint radiation known as the Cosmic Microwave Background represents the cooled remnant of the Big Bang itself. This ancient radiation fills the entire universe and provides one of the strongest pieces of evidence supporting the origin of the cosmos.
- **Infrared Radiation** - Infrared radiation lies just beyond the red end of visible light. Thermal cameras detect infrared radiation,

allowing us to visualise heat patterns in living organisms, machinery, or distant celestial bodies. Many astronomical objects that appear dark in visible light glow brightly in infrared wavelengths.

- **Visible Light** - The very obvious, it represents the narrow band of electromagnetic radiation detectable by the human eye.
- **Ultraviolet Radiation** - Beyond violet light lies ultraviolet radiation, which carries more energy than visible wavelengths. It can trigger chemical reactions in biological tissues, which is why prolonged exposure to sunlight can cause skin damage or sunburn.
- **X-rays** - X-rays possess significantly higher energies and shorter wavelengths than ultraviolet radiation. Because of their penetrating ability, X-rays are widely used in medical imaging to observe bones and internal structures within the body. They are produced when high-energy electrons collide with metal targets, releasing bursts of radiation capable of passing through soft tissue.
- **Gamma Rays** - At the most energetic end of the electromagnetic spectrum lie gamma rays. These are the most powerful electromagnetic waves known. They originate from nuclear reactions, radioactive decay, and some of the most violent events in the universe. Gamma-ray bursts, for instance, are among the brightest explosions ever observed, releasing enormous amounts of energy in mere seconds. Despite their destructive potential, gamma rays also reveal invaluable information about cosmic phenomena that would otherwise remain hidden.

The spectrum closes at the Gamma rays, and another category of radiation, which consists of actual particles moving at high speeds, starts. These particles may originate from radioactive decay, nuclear reactions, or cosmic events. What are they? The common types of particle radiation include:

- Alpha Particles - They are helium nuclei consisting of two protons and two neutrons, emitted during a certain type of radioactive decay.
- Beta Particles - They are high-speed electrons or positrons ejected from unstable atomic nuclei.
- Neutron Radiation - This occurs in nuclear reactions where neutrons escape from atomic structures.
- Cosmic Rays - They are high-energy particles originating from space, which constantly bombard the Earth's atmosphere, creating cascades of secondary particles that reach the surface.

Radiation is often classified according to whether it can ionise atoms. Ionising radiation carries enough energy to remove electrons from atoms, creating charged particles known as ions. Examples include:

X-rays

Gamma rays

High-energy ultraviolet radiation

Certain particle radiations

such as ionising radiation, can damage biological molecules such as DNA, which is why they must be handled with care. In contrast, non-ionising radiation lacks sufficient energy to ionise atoms. This category includes:

Radio waves

Microwaves

Infrared radiation

Visible light

. These forms of radiation are generally less harmful, though they can still interact with matter in significant ways.

Although radiation is often portrayed as mysterious or frightening, here are a few fascinating truths. Radiation surrounds us constantly. Natural sources include the Sun, cosmic rays, and radioactive elements within the Earth's crust. Beyond Earth's protective atmosphere, astronauts encounter intense radiation from solar particles and cosmic rays.

One of the most amusing examples of natural radiation involves bananas. Bananas contain potassium, and a tiny fraction of that potassium exists as the radioactive isotope potassium-40. As a result, eating a banana exposes the body to a minuscule dose of radiation. This dose is completely harmless but serves as an excellent reminder that radiation is not inherently unnatural.

We all know that radiation can be detected, but do you know how it can be seen? A curious demonstration of natural radiation can be seen using a cloud chamber. A Cloud chamber is a simple particle detector that allows us to visually observe the paths of charged particles. Inside the chamber, a supersaturated vapour of generally alcohol fills the air. When a high-energy particle passes through, it ionises the vapour along its path, causing tiny droplets to condense and form thin visible trails, almost like miniature contrails in the sky. Something very subtle yet fascinating would happen if we place a banana in it. Because bananas contain a small amount of the naturally radioactive isotope potassium-40, they emit occasional beta radiation. In the chamber, this would show up as faint, thin streaks appearing sporadically around the banana. Now, if a small chunk of Uranium were placed inside a Cloud chamber, the chamber would come alive with activity.

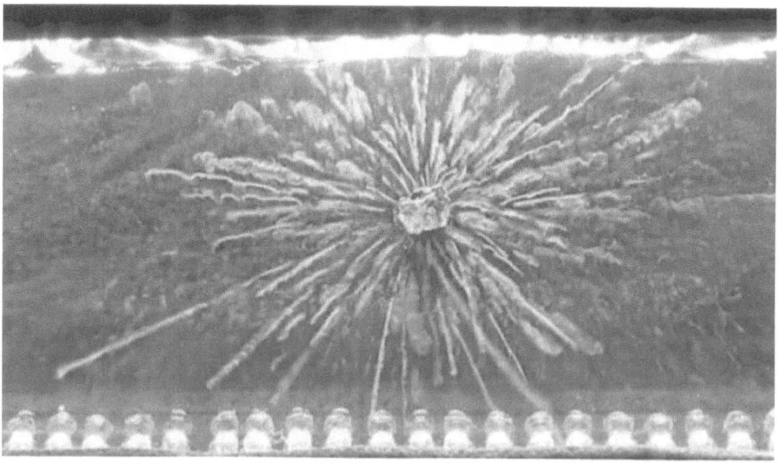

Figure 8.2: A Uranium Chunk in the Cloud Chamber
Source: IFLSCIENCE.com

Instead of occasional faint streaks, you would see numerous bright, thick tracks constantly appearing and branching out. It would look almost like a tiny storm of glowing trails, revealing the intense radiation emitted by the radioactive element.

Now, you would probably ask why we are studying radioactivity somewhat like in a 10th-grade Physics book, though it's different. It's because physics is not only a collection of equations and theories. It is also a history of experiments, discoveries and occasionally accidents, some moments where humans encountered nature in unexpected ways. So, to look at such stories, we first need to look at their core.

Tale of the Restless I

At the dawn of the twentieth century, the discovery of radioactivity had captivated the scientific world. In 1896, the physicist Henri Becquerel discovered that certain materials spontaneously emitted penetrating radiation. Soon afterwards, the very famous pioneering researchers Marie Curie and Pierre Curie isolated a new element that displayed extraordinary radioactive properties. This element was radium.

Radium fascinated both scientists and the public. It glowed faintly in the dark and emitted powerful radiation continuously without any visible source of energy. To many observers, radium seemed almost magical. During the early twentieth century, radium was widely regarded not as a dangerous substance but as a miracle material. It appeared in everything from medical treatments to consumer products. Radium-infused cosmetics, toothpastes, and even drinking water were marketed as health-enhancing innovations. The dangers of long-term radiation exposure were not yet fully understood. This misunderstanding would soon have devastating consequences.

During World War I and the years that followed, radium found an especially practical use. By mixing radium salts with phosphorescent compounds, manufacturers created a paint that glowed continuously in the dark. This luminous paint was ideal for marking the numbers and hands of watches, aircraft instruments, and military equipment. Such glow-in-the-dark dials allowed soldiers and pilots to read instruments at night without external lighting.

To produce these dials, factories hired workers, often young women, to carefully paint tiny numbers onto watch faces using extremely fine brushes. One of the most prominent facilities performing this work was located in Illinois, where a company known as the United States Radium Corporation operated a dial-painting factory. The workers who performed this delicate task would later become known as the

Radium Girls.

The work required extraordinary precision. Each dial contained tiny numerals and markers that had to be painted with luminous radium paint. To maintain sharp brush tips, workers were instructed to shape their brushes using a technique called "lip-pointing." After a few strokes, the paintbrush would lose its fine point. Workers were told to place the brush between their lips, roll it gently to sharpen it, and then dip it back into the radium paint. This process used to happen hundreds of times each day.

The women were assured that the paint was harmless. Workers sometimes painted their fingernails or teeth with the radium mixture, enjoying the eerie glow it produced under dim light. What they did not know was that every time they placed the brush in their mouths, they were ingesting tiny amounts of radioactive material.

Radium is a highly radioactive element. It decays by emitting several forms of radiation, including alpha particles, beta particles, and gamma rays. These emissions are forms of ionising radiation, which we discussed earlier in this chapter. Ionising radiation carries enough energy to remove electrons from atoms, creating ions and potentially damaging biological molecules.

When radium enters the human body, it behaves chemically in a particularly dangerous way. Radium closely resembles calcium, an element that plays a crucial role in bone formation. Because of this similarity, the body mistakenly incorporates radium into the skeletal structure. Once embedded in bone tissue, radium continues to decay, emitting radiation directly inside the body.

Several years after beginning their work, some of the dial painters began experiencing unusual health problems. Teeth began falling out. Jawbones developed painful infections. In some cases, sections of bone literally began to deteriorate. One particularly alarming condition became known as "radium jaw." In this disease, the bone tissue in the jaw would slowly decay and crumble. The damage resulted from the intense radiation being emitted by radium atoms lodged within the skeletal structure.

Now, this occurred due to radiation emitted by radium. But why? Because ionising radiation can damage living tissue in several ways.

When radiation passes through biological cells, it may:

Break chemical bonds in molecules.

Ionize atoms.

Disrupt DNA structures.

Generate highly reactive chemical species.

DNA damage is especially critical. When the genetic material inside cells is altered or broken, the cells may die or begin reproducing uncontrollably, potentially leading to cancer.

As more workers became ill, physicians and scientists began investigating the situation. One key researcher was the American physician Harrison Martland, who conducted detailed studies on the health effects of radium exposure. The findings confirmed that radium had accumulated in their bones and continued to emit radiation long after exposure. This discovery represented one of the earliest clear demonstrations of internal radiation poisoning.

The Radium Girls' case eventually led to legal battles between the workers and the companies responsible for their exposure. Their story deeply tragic. Many of the affected workers suffered severe illness, and several lost their lives. The tragedy ultimately contributed to the development of modern radiation protection standards. In a sense, we can say, their suffering illuminated the path toward safer scientific practices.

The Radium Girls experienced radiation through slow, invisible exposure to a radioactive substance. But there exists another story, perhaps even more astonishing, where radiation interacted with the human body in a far more dramatic and immediate way. Instead of radium paint, the source was a particle accelerator. Instead of gradual exposure, the radiation arrived in an instant. And somehow, against all expectations, the person involved survived.

Tale of the Restless II

Previously, we encountered the tragic story of the Radium Girls—an episode that revealed how invisible radiation, when absorbed by the body over time, could slowly damage living tissue. That story showed us radiation as a force that worked gradually within the body. But radiation does not always act slowly. Sometimes, it strikes in a single instant. Delivering energies that, according to our intuition, should be instantly fatal. And yet, to everybody's surprise, in one extraordinary case, a man survived such an encounter.

His name was Anatoli Bugorski. His experience remains one of the most astonishing and scientifically valuable accidents in the history of high-energy physics. To understand what happened to Bugorski, we must first understand the type of machine he was working with.

In the twentieth century, physicists began building enormous devices capable of accelerating subatomic particles to incredible speeds. These machines are known as particle accelerators. If you're thinking of those like the Large Hadron Collider, we discussed in Chapter 5. You're right.

We know, particle accelerators propel charged particles, such as electrons or protons, to velocities extremely close to the speed of light. Once accelerated, these particles can be smashed into targets or into each other, allowing scientists to probe the structure of matter at the smallest scales. Now, let us take a historical view of it too.

One such accelerator existed in the Soviet Union during the 1970s. It was called the U-70 synchrotron, located at the Institute for High Energy Physics in Protvino. This powerful device could accelerate protons to energies of 76 giga-electronvolts (GeV)—making it one of the most powerful particle accelerators in the world at that time. For comparison, a single proton inside this accelerator carried energy far greater than the kinetic energy of a microscopic particle in everyday conditions. When such particles travel in a beam, they form an intensely focused stream of high-energy matter. Under normal circumstances, no human being should ever come into contact with such a beam. But one day, in 1978, an extraordinary accident occurred.

At the time, Anatoli Bugorski was a young physicist working with the accelerator. One day, he was performing routine maintenance on part of the U-70 synchrotron system. According to reports, he believed the particle beam had been switched off. Unfortunately, it had not. As Bugorski leaned forward to inspect the equipment, his head accidentally crossed the path of the active proton beam. In that moment, the beam passed directly through his skull. The entry point was near the back of his head, and the exit point was near his nose. For a fraction of a second, an intensely concentrated stream of high-energy protons travelled through his brain tissue. In most circumstances, such exposure would be expected to cause immediate and catastrophic damage. Yet Bugorski did not lose consciousness. He later reported that he saw, "a flash brighter than a thousand suns", even though the accelerator tunnel was dark. Then the beam passed.

From a physics perspective, the proton beam in the accelerator was extremely unusual compared with natural radiation sources. Unlike the diffuse radiation from radioactive materials, a particle accelerator produces a highly focused beam. To get it, imagine a microscopic bullet made of protons, travelling at nearly the speed of light, concentrated into a narrow stream. Each proton carries enormous kinetic energy, and when such particles collide with matter, they can cause ionisation, atomic disruption, and secondary particle cascades. When the beam passed through Bugorski's head, it deposited radiation along a thin path through his brain. The energy was enormous—but the beam was incredibly narrow. This narrowness turned out to be the key reason Bugorski survived. Rather than destroying the entire brain, the beam damaged only a very thin channel of tissue. Still, the radiation dose in that path was estimated to be several thousand times higher than the dose normally considered lethal.

Soon after the accident, Bugorski began to experience symptoms. The side of his face where the beam entered began to swell. Over time, the skin along the beam path peeled away, revealing underlying tissue damage. The radiation had effectively burned a channel through his head. However, something remarkable happened. Instead of dying from massive brain damage, Bugorski survived. Doctors closely monitored his condition, expecting severe neurological decline. Instead, he gradually recovered from the initial injury. Though he suffered long-term effects—such as partial facial paralysis and hearing loss—his cognitive abilities remained largely intact. He was able to continue working as a scientist. And surprisingly, he is still alive as of 2026.

Let's consider the statement given by Bugorski about what he saw. Bugorski's report of seeing a brilliant flash during the accident offers another intriguing insight into radiation physics. Even though the accelerator tunnel was dark, he perceived an intense burst of light. We can say that this phenomenon likely occurred because the high-energy protons passing through his head interacted with the retina or nearby tissue, producing Cherenkov radiation or ionisation effects that stimulated the visual system. The brain interpreted this stimulation as light. But his survival does not mean the event was harmless. The radiation caused permanent physical damage, and the risks were immense. And this was the story of 'A Survivor of Particle Physics'.

The stories of the Radium Girls and Anatoli Bugorski show radiation interacting with matter—sometimes slowly, sometimes violently.

Lastly, we head next to some theoretical models that suggest the existence of exotic cosmic structures where matter and spacetime might behave in entirely unfamiliar ways. Among these possibilities are objects known as white holes, the hypothetical opposites of black holes, and tunnels through spacetime called wormholes.

The Spacetime Spillway

Wormholes

Picture spacetime not as a static stage upon which the universe performs its drama, but as a restless river—flowing, bending, and curving under the influence of mass and energy. In the framework of modern physics, especially the theory of general relativity, we know that gravity is not a force pulling objects together through empty space. Instead, gravity is the shape of spacetime itself. Planets create gentle dimples in this cosmic fabric. Stars carve deeper wells. Black holes twist spacetime into extreme curvatures from which even light struggles to escape.

Now imagine that this vast cosmic river sometimes behaves in even stranger ways. Imagine places where the fabric folds over itself, forming hidden shortcuts between distant regions of space. Imagine locations where spacetime itself appears to burst outward, releasing matter and energy into the universe. This conceptual landscape can be poetically described as the Spacetime Spillway. Within this strange landscape lie two fascinating ideas: wormholes and white holes. Both arise naturally from the mathematics of general relativity, and both stretch our understanding of space, time, and cosmic structure. So, let's begin!

The concept of wormholes first emerged from the mathematics of Einstein's gravitational equations. In 1935, the physicists Albert Einstein and Nathan Rosen examined solutions describing black holes and discovered a curious feature hidden within the equations. Under certain circumstances, spacetime could form a bridge connecting two distant regions. This structure became known as the Einstein–Rosen bridge. Today, we commonly call such structures wormholes.

To picture a wormhole, imagine the universe as a flat sheet of paper. Two points on that sheet may be separated by a great distance. But if the sheet is folded so that the points touch, a tunnel through the fold would connect them directly. A wormhole acts much like that tunnel. Instead of travelling the long distance through normal spacetime, an object could theoretically pass through the wormhole and emerge somewhere else entirely—possibly thousands or even millions of light-years away. This idea has captured the imagination of both scientists and storytellers alike.

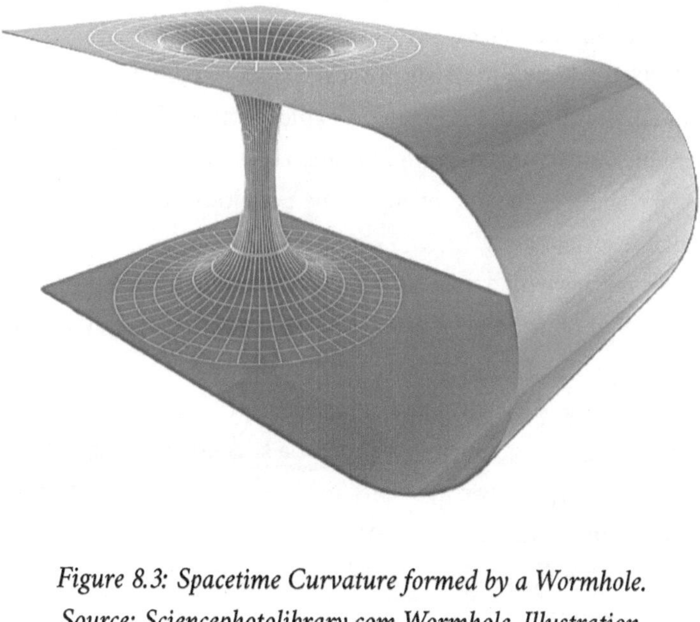

Figure 8.3: Spacetime Curvature formed by a Wormhole.
Source: Sciencephotolibrary.com Wormhole, Illustration

If you have seen the remarkable film *Interstellar* directed by Christopher Nolan, you may already have a visual sense of how a wormhole might appear: a spherical distortion in space that bends light from the surrounding stars. Although the cinematic representation simplifies many aspects of physics, the core concept is grounded in genuine theoretical work. So, you may wonder, if the math works, why is this theoretical?

Early mathematical models of wormholes revealed a major obstacle. Most wormholes predicted by general relativity would collapse almost immediately after forming. The intense gravitational forces around the wormhole's throat—the narrow passage connecting two regions of spacetime—would cause the tunnel to pinch shut before anything could pass through it. In other words, wormholes might exist mathematically, but they would be non-traversable.

In 1988, however, physicists proposed a new possibility. Researchers Michael Morris and Kip Thorne explored the idea of traversable wormholes, structures that might remain open long enough for matter, or even travellers, to pass through. Their solution required something unusual: exotic matter.

Exotic matter would possess negative energy density, a property opposite to that of ordinary matter. Instead of attracting gravitationally, it could create repulsive effects that prevent the wormhole's throat from collapsing. Though such matter has never been observed in large quantities, certain quantum phenomena suggest that small amounts of negative energy may exist. This 'may' right here is why wormholes exist mathematically but may not exist physically. On top of that, whether such effects could ever be scaled up to stabilise a wormhole remains unknown.

But there is one more intriguing idea about wormholes that was proposed by the physicist John Archibald Wheeler. So what was it? Wheeler suggested that at extremely small scales, close to the Planck length, spacetime might not be smooth and continuous at all. Instead, it could resemble a constantly fluctuating environment known as quantum foam, which I've mentioned previously a few times.

In this picture, spacetime continually forms and destroys microscopic tunnels, loops, and distortions. Among these fleeting structures could be tiny wormholes, appearing and disappearing within unimaginably short intervals. These wormholes would be far too small and unstable for travel, but their existence would hint that spacetime possesses a deeply dynamic and interconnected structure. This is quite what we can say about wormholes for now.

White Holes

If wormholes represent hidden tunnels within spacetime, white holes represent something even more unusual. A white hole can be thought of as the time-reversed counterpart of a black hole. What black holes are is something which we certainly know. They pull matter inward with overwhelming gravitational force, trapping anything that crosses their event horizon.

White holes, in contrast, would behave oppositely. Nothing could enter a white hole. Instead, matter and energy would continuously emerge from it, expelled outward into the surrounding universe. In the mathematical solutions of Einstein's gravitational equations (particularly the Schwarzschild solution describing non-rotating black holes), white holes appear naturally as mirror structures in spacetime.

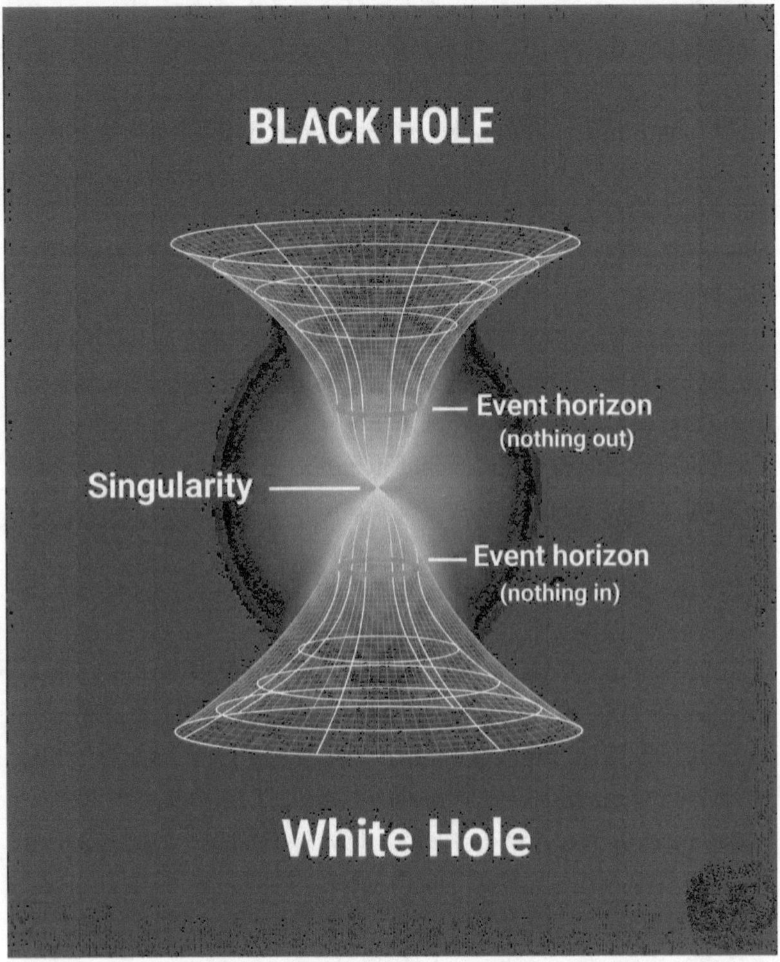

Figure 8.4: The concept of White Hole.

However, whether such objects can physically exist in our universe remains uncertain. Or, I can say very uncertain. Why? Unlike black holes, which have been observed indirectly through their gravitational influence and radiation signatures, no confirmed white hole has ever been detected. Despite the lack of observational evidence, white holes remain an intriguing possibility in theoretical physics.

Some cosmological ideas suggest that the Big Bang itself might resemble a white hole event. A moment when matter and energy erupted outward into an expanding universe. In other theoretical models, black holes may not end in singularities at all. Instead, the collapse of matter might eventually halt due to quantum effects, leading to a dramatic rebound or cosmic bounce. In such scenarios, a black hole could eventually transform into a white hole, releasing the matter it once consumed.

These ideas appear in certain approaches to quantum gravity, where spacetime is believed to possess a discrete structure rather than a perfectly smooth continuum. If correct, such theories might resolve the paradoxes surrounding black hole singularities.

In the end, I would say that the study of spacetime is not merely about geometry. It is about discovering how the universe itself is woven and whether hidden passages may exist within its grand design.

Denouement VIII

Every journey through physics leaves behind a trail of questions—some answered, many transformed, and a few that become even more mysterious the deeper we explore. This chapter has been such a journey. It has wandered through the quieter corners of physics, pausing at ideas and stories that did not fit neatly into earlier discussions but still hold profound meaning in our attempt to understand the universe.

We began with a curious thought experiment: the Apple in a Box. At first glance, it seemed like a simple scenario, but it turned out to reveal astonishing facts. From that quiet thought experiment, our exploration moved into a more energetic domain—the restless world of radiation.

To illustrate this, we briefly stepped into the world of experimental detection with the cloud chamber. The story of the Radium Girls in Illinois. The extraordinary human encounter with radiation of Anatoli Bugorski. From there, the chapter expanded outward toward some of the most exotic ideas predicted by modern physics. We entered the strange territory of wormholes and closely related white holes.

Before we move on to the finale chapter of the book, I want to say something. The next chapter is different from any other in this book; it holds something which just did not come into my mind like Eureka! But a compilation of thoughts while in my journey of standard and theoretical physics. What is singularity? How did cosmogenesis occur?

Meet you in the next chapter...

9

Abstract Conduit

"RW5lcmd5"

There are questions in physics that feel like distant mountains on the horizon. We know they exist. We know they are important. Yet every time we approach them, the path seems to split into new uncertainties. Among the greatest of these questions is one that stands quietly at the centre of modern cosmology: What exactly is a singularity?

In the language of physics, a singularity is a point where our equations break down. It is where density appears to become infinite, spacetime curvature diverges, and the known laws of physics lose their predictive power. According to general relativity, singularities lie at the heart of black holes and at the very beginning of our universe during the Big Bang. But there is a problem. While our equations point toward singularities, no one truly knows what they are. Are they real physical objects? Are they merely mathematical artefacts that signal the failure of our current theories? Or are they gateways to deeper laws of nature that we have not yet discovered? These questions remain open.

Even the greatest physicists have struggled with them. The mathematics of singularities was deeply explored by thinkers like Roger Penrose and Stephen Hawking, who showed that under reasonable conditions, singularities must form within collapsing stars and within the early universe itself. Yet their work also revealed a troubling truth: general relativity predicts singularities, but it cannot describe what happens inside them. To understand that region, we would likely need a complete theory of quantum gravity, a framework that unifies the geometry of spacetime with the probabilistic laws of quantum mechanics. Until such a theory emerges, singularities remain among the greatest mysteries in physics. But there is another way to approach such mysteries. If our equations cannot yet tell us everything, perhaps imagination can guide us toward new ways of thinking about them. So let us attempt something unusual.

Let us imagine that we could visit a singularity. Not merely observe it from afar, but approach it, study it, and perhaps even cross the threshold where spacetime itself bends beyond recognition. Of course, we know, such a journey presents a serious obstacle - Black Holes. Black holes are among the most hostile environments in the universe. Their gravity stretches objects into thin streams of matter, a process sometimes called 'spaghettification'. Their immense tidal forces can tear apart atoms long before any spacecraft reaches the central region. Even light struggles to escape their grip. Clearly, if we are to attempt such a voyage, even in thought, we must begin with a vehicle unlike anything humanity has ever built. We must design a spaceship capable of travelling into the most extreme environments known to physics. But before we begin constructing such a craft, an important declaration must be made.

The spaceship we are about to design is completely fictional. Human technology today cannot harness gravity as an engineering tool, cannot manipulate the structure of spacetime, and certainly cannot work directly with stellar cores or quantum gravitational phenomena. However, the principles we will use in imagining this spacecraft will be rooted in real scientific ideas. The components of this vessel will draw inspiration from genuine physical concepts, such as gravitational fields, relativistic effects, nuclear processes, and spacetime geometry.

Think of this as a scientific thought experiment. If some sufficiently advanced beings, perhaps millions or billions of years ahead of us in technological evolution, were capable of harnessing gravity, controlling nuclear energies, and shaping spacetime itself, then a spacecraft constructed with such principles might not be entirely impossible. In that sense, our journey will be fictional in engineering but honest in physics.

So let us begin. And that means it is time to start the construction of our spacecraft.

Craftsmanship

In this spirit, let us imagine an advanced intelligence, something far beyond human technological capability, capable of manipulating the deepest forces of nature. Let us call this hypothetical architect 'The Weaver', a collective intelligence of ultra-precise particle engineers capable of shaping matter at the most fundamental scales. To attempt such a journey, we need to ask The Weaver to construct a spacecraft unlike anything humanity has ever conceived. He agrees, and the construction starts.

The vessel is named 'Voidfarer'. Everything about the Voidfarer is fictional in engineering, yet grounded in real physical principles. Each component draws inspiration from concepts already present in modern theoretical physics.

The Core

Every spacecraft begins with a source of energy. For a mission toward a black hole, conventional fuel systems would be hopelessly inadequate. The Weaver, therefore, constructs a compact quantum reactor at the heart of the Voidfarer.

The reactor relies on a phenomenon known as the 'Casimir effect', a subtle but experimentally verified consequence of quantum mechanics. When two extremely close conducting plates are placed a tiny distance apart, quantum fluctuations in the vacuum create a small attractive force between them. This occurs because certain wavelengths of virtual particles cannot exist between the plates, producing a difference in energy density between the inside and outside regions. In simple terms, empty space itself contains energy.

By arranging countless microscopic plate structures like an intricate lattice of squeezed mirrors, the Weaver amplifies this effect to generate controlled regions of negative energy density. This Casimir reactor becomes the heart of the Voidfarer, providing a continuous and stable energy supply without the need for conventional fuel. To support the reactor, the core is housed within a structural skeleton composed of graphene-based frameworks. Graphene, an ultra-thin sheet of carbon atoms arranged in a hexagonal lattice, is among the strongest materials ever studied. Woven together into layered structures, it forms a lightweight yet remarkably durable support system for the ship's internal machinery.

The Hull

. The survival of the Voidfarer depends entirely on the strength of its outer shell. To construct it, the Weaver uses one of the strangest materials predicted by astrophysics: nuclear pasta.

Deep within the crust of neutron stars, matter exists in exotic configurations unlike anything found on Earth. Under immense gravitational pressure, protons and neutrons become packed so tightly that conventional atomic structure disappears. Instead of individual nuclei surrounded by empty space, the particles arrange themselves into unusual geometric patterns resembling strands, sheets, and tubes. Because these shapes resemble different types of pasta, physicists humorously named them nuclear pasta.

In this state, matter becomes almost unimaginably dense, around:

10^{14} grams per cubic centimeter

and its structural strength is estimated to be billions of times greater than steel.

The Weaver shapes this material into a 20 metre egg-shaped hull surrounding the Voidfarer. The shell consists of thin-layered slabs of nuclear pasta arranged like armour plating. These slabs resemble the "lasagna phase" of nuclear pasta. They are flat, stacked sheets of degenerate nuclear matter. So, why is this so powerful? Because inside nuclear pasta, there is essentially no empty space. The neutrons and protons occupy every available quantum state, forming what physicists call a degenerate matter configuration. In such a state, particles cannot simply move aside or compress further. And this produces several remarkable properties, the reason why we are using this material:

Extreme Strength – The tightly packed particles resist deformation, making the structure incredibly resistant to crushing forces.

Radiation Resistance – High-energy particles such as gamma rays or cosmic rays cannot easily penetrate the material. Instead, their energy is absorbed or dispersed through the dense lattice.

Thermal Stability – The degenerate packing of particles prevents large temperature fluctuations from significantly altering the structure.

In short, nuclear pasta is perhaps the closest thing the universe offers to indestructible armour. However, one enormous challenge remains.

Outside a neutron star, such matter would normally explode apart instantly, because the immense pressure holding it together would disappear. To prevent this, the Weaver uses the negative-energy fields generated by the Casimir reactor. These fields mimic the crushing gravitational pressure found inside neutron stars, stabilising the nuclear pasta hull even in ordinary space.

The Travel Method

Let's aim for a supermassive black hole. Why not none other than our Sagittarius A*, residing at the centre of our Milky Way. But reaching a black hole such as Sagittarius A* would normally require travelling about 26,000 light-years. Even at near-light speeds, such a journey would take tens of thousands of years. And this is why the word 'normally' is used. The Weaver solves this problem not by building faster engines, but by changing the geometry of space itself.

Surrounding the ship's core are rings composed of negative-energy filaments. These rings manipulate spacetime in a way inspired by the mathematics of Einstein–Rosen bridges. In essence, the rings create and stabilise a wormhole. One end of the wormhole remains near the ship's launch location. The other opens near Sagittarius A*, instantly connecting two vastly distant regions of space. A slight rotation of the core's energy field folds spacetime inward, opening the tunnel. Once formed, the Voidfarer simply drifts through the passage. For the crew, the trip lasts only seconds. Maintaining the wormholes requires exotic forms of negative energy, precisely what the Casimir reactor was designed to produce.

Gravitational Protection

Now, approaching a black hole introduces another terrifying danger: tidal forces. Gravity near a black hole is not uniform. The gravitational pull on one side of an object may be dramatically stronger than on the other side. This difference stretches objects into long strands, a phenomenon often described as spaghettification, not the way spaghetti is made. To counter this, the Voidfarer is equipped with gravitational dampers.

These devices use carefully tuned magnetic and quantum field configurations to counteract tidal forces. The system draws inspiration from a real relativistic effect known as frame dragging, in which rotating massive objects twist the surrounding spacetime. By manipulating these effects, the dampeners create controlled counter-gradients that reduce the stretching forces acting on the ship. The nuclear pasta hull plays an important role as well. Because the material barely flexes under stress, it acts as an almost perfect structural stabiliser. Together, the dampeners and hull allow the ship to maintain structural integrity even under extreme gravitational conditions.

The Pilot Module & Journey

Obviously, we need to observe what is happening with the spaceship and the surroundings. Most importantly, the singularity. So, to make it possible, inside the Voidfarer lies a small central chamber: the pilot pod. The mission does not necessarily require a biological human pilot. Instead, the Weaver designs the cockpit to accommodate either a digitised consciousness or a human operator connected through neural interfaces.

Surrounding the pod are advanced sensors capable of detecting phenomena invisible to ordinary instruments. These include: Neutrino detectors, allowing the ship to observe particles that pass through almost all matter. Gravitational wave sensors are capable of mapping distortions in spacetime itself. Radiation scanners are monitoring high-energy particle fields near the black hole. Because visible light becomes increasingly distorted near the event horizon, these instruments reconstruct the environment through computational visualisation, feeding the information directly into the pilot's perception. Even in total darkness, the pilot can see the geometry of spacetime itself.

Now, be ready, as with its systems active, the Voidfarer begins its descent toward the black hole. From orbit, the wormhole system activates, and the ship emerges near Sagittarius A*. As the ship approaches the event horizon, time dilation becomes extreme. To distant observers, the Voidfarer appears to slow almost to a halt. But inside the ship, the journey continues normally. The horizon is crossed.

Contrary to popular imagination, nothing dramatic marks the crossing itself. No wall or barrier appears. The ship simply drifts inward into a region where all paths lead toward the centre.

As the Voidfarer moves deeper into the black hole, its instruments deploy tiny entangled probes toward the central region. According to classical general relativity, the singularity should be a point of infinite density—a location where the known laws of physics break down. But the data returned by the probes suggests something different. The probes then sense the data and records them. After gathering its data, the Voidfarer reopens its wormhole tunnel and returns. The recorded data suggests something unexplainably astonishing.

Pre-Conjecture

Before we move any further, it is important to pause for a moment and look back at what we have just experienced. The journey of the Voidfarer, its construction, its exotic hull of nuclear pasta, its wormhole passage across the galaxy, and its descent toward the heart of a black hole, was never meant to be taken as a literal engineering proposal. It was a semi-fictional voyage, a carefully crafted thought experiment. Every component of the ship borrowed ideas from real physics, yet the combination of those ideas stretched far beyond our current technological reach. In other words, what you have just read was not the main argument of this chapter. It was an intellectual warm-up. The Voidfarer was simply a vehicle of imagination. Its mission was to take us mentally to the place where our current understanding of physics reaches its limits.

By picturing a spacecraft capable of reaching such an extreme region, even if only hypothetically, we allow ourselves to ask questions that ordinary descriptions cannot easily address. What would the singularity actually look like? Would it truly be a point of infinite density? Would it be replaced by some quantum structure? Could information survive there, or would it vanish completely? We used the Voidfarer only as a narrative device. Because the true subject of this chapter has not yet been revealed.

What this chapter is ultimately building toward is a conjecture about singularity. A conjecture about cosmogenesis. A conjecture defining the probability of multiverses. Now, a conjecture is not the same as a theory.

In science, a theory represents a framework supported by extensive mathematical formulation and strong empirical evidence. Theories such as general relativity or quantum mechanics have survived countless experimental tests and form the backbone of modern physics.

A conjecture, on the other hand, is something more modest yet often just as important. A conjecture is an informed proposal—a logical idea suggested by existing knowledge but not yet proven. It may arise from patterns, intuition, or philosophical reasoning within the structure of known physics. Sometimes conjectures eventually evolve into fully developed theories. At other times, they inspire new experiments or entirely new ways of thinking. In short, a conjecture is a seed. It is a possibility waiting for evidence.

The Voidfarer has completed its imaginary journey. Now the real destination awaits. In the next and final segment of this book, we will examine the conjecture itself. A considerable thing you can expect is the use of a very formal language ahead; I just wanted to make it feel professional.

The Ashmit Cosmogenic Conjecture

A Hypothesis for Black Hole–Induced Cosmogenesis

Abstract

The **Ashmit Cosmogenic Conjecture** proposes that extreme gravitational collapse within black holes may not culminate in classical singularities but instead initiate the formation of new spacetime regions. As spacetime curvature approaches Planck-scale magnitudes, quantum gravitational effects are expected to become dominant, potentially inducing a topological transition that connects the black hole interior to a newly nucleated universe. Matter and information entering the black hole may then contribute to the physical state of this emergent spacetime, allowing for global information preservation and offering a possible resolution to the black hole information paradox. The conjecture further suggests that universes may exist as layered spacetime manifolds embedded within a higher-dimensional structure, remaining causally disconnected under normal conditions but capable of transient interaction during extreme gravitational events. Although speculative, this framework establishes a conceptual bridge between black hole physics, cosmology, and quantum gravity and provides a basis for exploring the cosmogenic role of gravitational collapse.

Motivation

According to General Relativity, sufficiently massive stars collapse under gravity to form black holes. These objects are bounded by an event horizon beyond which no information can escape. At the centre of the black hole, classical theory predicts a singularity, a point at which spacetime curvature becomes infinite, and the laws of physics break down.

Many physicists interpret such infinities as indicators that classical gravitational theory is incomplete. It is widely expected that a complete theory of quantum gravity will replace singularities with a finite physical mechanism.

The Ashmit Cosmogenic Conjecture explores the possibility that extreme curvature inside black holes may produce not a singularity but a transition in spacetime structure capable of generating a new universe, consequently getting called a singularity.

Conceptual Framework

Proposed Topological Rupture

At the core of the conjecture lies the idea that gravitational collapse may lead not to a breakdown of spacetime, but to a transformation of its structure. The mechanism underlying this process is described as a Planck-Scale Topological Transition (PSTT).

As gravitational collapse proceeds, matter and energy become increasingly compressed within the black hole's interior. The local energy density rises toward a critical Planck-scale threshold, while spacetime curvature simultaneously approaches its maximum physically sustainable value. At this stage, the predictions of classical General Relativity, which is governed by the Einstein Field Equations, become incomplete, and quantum gravitational effects are expected to dominate.

Near this limit, spacetime is hypothesised to enter a highly fluctuating regime often described as 'quantum foam', a concept originally suggested by John Wheeler. In this regime, spacetime loses its smooth manifold structure and becomes dynamically unstable at the Planck scale.

The PSTT is triggered when both critical thresholds are approached simultaneously. At this point, instead of forming a classical singularity, spacetime undergoes a topological transition, which is a fundamental change in its connectivity. This transition results in the formation of a temporary bridge between our universe, M_A and a newly nucleated spacetime region M_B.

Matter and energy crossing this region are redirected into the emergent spacetime, contributing to its formation and early expansion. In this sense, the black hole interior acts not as a terminal point, but as a cosmogenic gateway.

Layered Universe Hypothesis

Within this framework, universes may be understood as distinct spacetime manifolds embedded within a larger structure. Under ordinary conditions, these manifolds remain causally disconnected. However, extreme gravitational environments—such as those found in black holes—may enable transient connections between them.

This layered picture bears conceptual similarity to higher-dimensional cosmological models, including ideas explored in brane cosmology.

It also shares a philosophical resemblance to Cosmic Natural Selection, proposed by Lee Smolin, in which black holes give rise to new universes with varying properties. However, the present conjecture differs in its emphasis on topological transitions and information transfer mechanisms. And I came to know of it after the drafting of the conjecture.

Quantum Entanglement & Information Transfer

Information transfer across the cosmogenic transition can be understood through quantum entanglement between the black hole interior and the Hawking radiation emitted at the event horizon.

As the system approaches a critical curvature, the quantum states of infalling matter become strongly entangled with the outgoing radiation. This creates a non-local connection in which the internal state of the black hole is correlated with the radiation escaping from it.

Because of entanglement, the evolution of the total system remains unitary, meaning information is conserved globally. This entanglement can be thought of as a kind of 'quantum mirroring'. To understand this, imagine a pair of perfectly synchronised dice: One die falls into the black hole. The other remains outside. Even though you can no longer see the one inside, the outcome of the outside die still carries information about it.

Physically, a part of the information goes into the new universe M_B. A part is encoded in Hawking radiation. This represents a "Best of Both Worlds" resolution to the black hole information paradox identified by Stephen Hawking, where information is globally preserved even if it appears locally inaccessible.

Additionally, the entanglement structure between connected spacetime regions is suggestive of ideas related to the Maldacena Conjecture, where spacetime geometry and quantum entanglement are deeply interconnected.

Hawking Radiation and Black Hole Evaporation

Stephen Hawking discovered that black holes emit radiation due to quantum effects. Over extremely long times, this radiation causes black holes to shrink. The topological transition initiated by the

PSTT does not produce a permanent structure. Instead, the resulting spacetime connection persists dynamically throughout the lifetime of the black hole. The topological transition initiated by the PSTT does not produce a permanent structure. Instead, the resulting spacetime connection persists dynamically throughout the lifetime of the black hole.

As evaporation proceeds, the bridge contracts continuously. When the black hole fully evaporates, the core conditions fall below the critical thresholds, and the topological connection closes naturally. This ensures that matter and entangled information can flow into the emergent universe throughout the entire lifetime of the black hole, with the connection terminating precisely at the final stage of evaporation.

Implications

If the conjecture were correct, several implications would follow-

First, classical singularities would be replaced by finite quantum gravitational processes.

Second, information entering black holes would not be destroyed but transferred to the quantum state of a newly formed universe.

Third, black holes could serve as mechanisms through which universes generate new universes, potentially forming a cosmological network of connected spacetime regions.

Fourth, the multiverse could evolve through continuous gravitational collapse and cosmogenesis.

Finally, our own universe might itself have originated through a similar process within a black hole of a preceding universe.

Postulates of the Conjecture

Axiom I - Extreme Curvature Limit

When gravitational collapse within a black hole drives spacetime curvature toward Planck-scale magnitudes, classical descriptions based on General Relativity become incomplete. In this regime, quantum gravitational effects are expected to dominate, preventing the formation of physically ill-defined infinities. This axiom asserts that singularities predicted by classical theory signal the breakdown of the theory itself, rather than the existence of true physical infinities, and must therefore be replaced by a finite quantum-gravitational description.

Axiom II - Topological Transition of Spacetime

At or near the Planck-scale curvature limit, spacetime may undergo a topological transition rather than collapsing into a classical singularity. This transition alters the connectivity of the spacetime manifold, resulting in the formation of a new spacetime region connected to the black hole interior. This process establishes a temporary but physically meaningful connection between two otherwise disconnected spacetime domains, transforming the classical endpoint of collapse into a cosmogenic transition.

Axiom III - Layered Manifold Structure

The cosmos may consist of multiple, dynamically independent spacetime manifolds embedded within a higher-dimensional structure. Each manifold (universe) evolves according to its own internal physical laws and remains causally disconnected under normal conditions. However, under extreme gravitational environments—such as those present within black holes—temporary connections between these manifolds may form. This layered structure is conceptually aligned with ideas explored in higher-dimensional cosmological models and brane-like frameworks.v

Axiom IV - Conservation of Information Across Universes

Information carried by matter and radiation entering a black hole is not destroyed but redistributed across a larger quantum system that includes both the parent universe and the emergent spacetime region.

Through quantum entanglement, information becomes encoded in a correlated manner between the newly formed universe and the radiation emitted from the black hole.

Thus, while information may appear inaccessible from a local perspective, it remains universally conserved, offering a possible resolution to the black hole information paradox first highlighted by Stephen Hawking.

Axiom V - Evaporation and Closure of the Rupture

Black holes undergo gradual mass loss through quantum processes, most notably Hawking radiation. As the black hole evaporates, the physical conditions required to sustain the topological transition weaken. The spacetime connection between universes persists throughout the lifetime of the black hole but continuously diminishes as its mass decreases. Upon complete evaporation, the conditions necessary for the transition cease to exist, and the connection closes naturally.

Conclusion

The Ashmit Cosmogenic Conjecture proposes that extreme gravitational collapse may lead to the formation of new universes rather than classical singularities. By linking black hole physics, quantum information preservation, and cosmogenesis, the conjecture offers a speculative perspective on the deeper structure of spacetime and the possible evolutionary dynamics of the cosmos.

Future developments in gravitational physics, cosmology, and quantum theory may determine whether such ideas correspond to physical reality.

Culmination

With all that, we culminate the conjecture. The conjecture does not claim certainty. It does not stand as a completed theory. Instead, it offers a speculative framework suggesting that singularities may not be destructive endpoints of spacetime but rather transitional regions, where the extreme conditions of gravity and quantum physics give rise to new cosmic domains.

At this point, let me tell you something: if such a view were even partially correct, it would imply that the formation of a black hole in one universe might correspond to the birth of another universe elsewhere. The death of stars could quietly seed entirely new cosmic structures beyond our observational reach. In other words, singularities might not be the end of spacetime at all.

Now, if singularities can connect regions of spacetime in such profound ways, then another enormous question naturally follows:

Could they connect entire universes?

This brings us to the concept of the multiverse, the possibility that our universe may be only one among many. But that is a discussion for another time. Perhaps even for another book.

For now, this journey reaches its conclusion...

Technical Formulation

Appendix A - Mathematical Background

This section is probably not for you if you don't want math-heavy pages. Consider this optional, but if you want to take a look at the technical depth of the conjecture. Do take a look.

Extreme Curvature at the Black Hole Core

Spacetime curvature near a black hole can be measured using the Kretschmann Scalar:

$K = 48G^2M^2 / c^4r^6$

Where:

K = curvature of spacetime

G = gravitational constant

M = mass of the black hole

c = speed of light

r = distance from the centre of the black hole

As r approaches zero, curvature approaches infinity in classical physics. This indicates that the theory becomes incomplete under such extreme conditions.

Planck Scale and Quantum Gravity

Physics suggests that there exists a fundamental length scale where quantum gravitational effects dominate. This is known as the Planck Length:

$lp = \sqrt{(\hbar G / c^3)}$

Where:

lp = Planck length (1.616×10^{-35} m)

$\hbar$ = reduced Planck constant

G = gravitational constant

c = speed of light

At this scale, spacetime may behave very differently from classical expectations.

Proposed Topological Rupture Limit (Hypothetical)

The PSTT is triggered when both critical thresholds are approached simultaneously:

$Kmax = 1 / (l_p)^2$

Where:

$Kmax$ = maximum spacetime curvature

l_p = Planck length

$Kmax \approx 3.828073 \times 10^{69}\ m^{-2}$

$Kcrit = 1 / (l_p)^4$

Where:

$Kcrit$ = critical energy density

lp = Planck length

$Kcrit \approx 1.465414 \times 10^{139}\ m^{-4}$ (Numerically)

$Kcrit \approx 4.63295 \times 10^{113}\ J/m^3$ (SI Energy-density units)

These thresholds define the regime in which classical spacetime is expected to become unstable, potentially triggering the proposed topological transition.

Universal Representation

The conjecture imagines that multiple universes may exist as separate spacetime manifolds:

$M_A, M_B, M_C \ldots$

Where:

M_A = our universe

M_B = a newly formed universe

M_C = another possible universe

Each universe follows the Einstein Field Equations:

$G\mu\nu = 8\pi GT\mu\nu$

Where:

$G\mu\nu$ = spacetime curvature tensor

$T\mu\nu$ = energy–momentum tensor

G = gravitational constant

μ, ν = spacetime coordinate indices

Information Transfer

This can be described in quantum terms using Hilbert spaces:

$H_{total} = H_A \otimes H_B$

Where:

H_{total} = total quantum state space

H_A = quantum state space of our universe

H_B = quantum state space of the new universe

$\otimes$ = tensor product combining both systems

This formulation allows for a unitary description of the combined system.

Hawking Radiation Temperature

The temperature of this radiation is:

$T_H = \hbar c^3 / 8\pi G M k_B$

Where:

T_H = Hawking temperature

$\hbar$ = reduced Planck constant

c = speed of light

G = gravitational constant

M = mass of the black hole

k_B = Boltzmann constant

This relation governs the evaporation of black holes and, within the conjecture, the evolution of the cosmogenic transition.

Appendix B - Topological Rupture Derivation

Below are the three complete, step-by-step derivations presented together for clarity and consistency. All derivations start from the fundamental constants *c*, *G*, and ħ, and connect directly to the reduced Planck length l_P.

1. Derivation of Kmax (Maximum Spacetime Curvature)

Using Dimensional Analysis

Step 1: Dimensions of constants

- c: $[L^1 T^{-1}]$
- G: $[L^3 M^{-1} T^{-2}]$
- ħ: $[M^1 L^2 T^{-1}]$

Step 2: Assume $K_{max} = c^x G^y \hbar^z$ and set equal to curvature units $[L^{-2} M^0 T^0]$:

$(L^1 T^{-1})^x \cdot (L^3 M^{-1} T^{-2})^y \cdot (M^1 L^2 T^{-1})^z = L^{-2} M^0 T^0$

Step 3: System of equations

- Length (L): $x + 3y + 2z = -2$
- Mass (M): $-y + z = 0$
- Time (T): $-x - 2y - z = 0$

Step 4: Solve from M: $z = y$
Substitute into T: $-x - 3y = 0 \Longrightarrow x = -3y$
Substitute into L: $-3y + 3y + 2y = -2 \Longrightarrow 2y = -2 \Longrightarrow y = -1$
Thus $z = -1$, $x = 3$.

Step 5: Final expression
$K_{max} = c^3 G^{-1} \hbar^{-1} = c^3 / \hbar G$

Step 6: Link to Planck length
$l_p^2 = \hbar G/c^3 \Longrightarrow K_{max} = 1/l_p^2$

Numerical value: $K_{max} \approx 3.828073 \times 10^{69}\ m^{-2}$

2. Derivation of Kcrit in Maximum Energy Density – Geometric / Natural Units

Direct algebraic substitution

Step 1: Reduced Planck length

$l_p = \sqrt{\hbar G c^3}$

Step 2: Raise to the fourth power

$l_p{}^4 = (\hbar G / c^3)^2 = \hbar^2 G^2 c^6$

Step 3: Reciprocal

$1 / l_p{}^4 = c^6 \hbar^2 G^2$

Step 4: In natural units ($\hbar = c = 1$)

$K_{crit} = 1 / l_p{}^4$

Numerical value (in m^{-4}):

$K_{crit} \approx 1.465414 \times 10^{139}\ m^{-4}$

3. Derivation of Kcrit (Maximum Energy Density – Physical Form) Using Planck

Energy divided by Planck Volume

Step 1: Planck mass

$m_p = \sqrt{\hbar c / G} = \hbar / c\, l_p$

Step 2: Planck energy

$E_p = m_P c^2$ (from $E = mc^2$) $= \hbar c / l_P$

Step 3: Planck volume

$V_P = l_P^3$

Step 4: Critical energy density

$K_{crit} = E_P / V_P = (\hbar c / l_P) / l_P^3 = \hbar c / l_P^4$

Step 5: In natural units ($\hbar = c = 1$)

$K_{crit} = 1 / l_P^4$

Numerical value (SI energy-density units):

$K_{crit} \approx 4.63295 \times 10^{113}\ \mathrm{J/m^3}$

Appendix C - Falsifiability

Although the Ashmit Cosmogenic Conjecture concerns processes occurring within black holes—regions that cannot be directly observed—there may still exist indirect observational consequences. If future observations contradict these predictions, the conjecture would be falsified.

Information Preservation in Black Hole Evaporation

The conjecture predicts that information is not destroyed but encoded in Hawking radiation through quantum correlations. If Hawking radiation is experimentally or theoretically confirmed to be perfectly thermal and completely uncorrelated, with no mechanism for encoding information, then the assumption of global information preservation would be invalidated.

Conversely, the detection of subtle correlations or deviations from perfect thermality would support the conjecture and align with attempts to resolve the information paradox identified by Stephen Hawking.

Final Stage of Black Hole Evaporation

The conjecture predicts that the connection between universes closes when the black hole completely evaporates. This implies that the final stages of black hole evaporation may show non-classical radiation signatures. Future observations of microscopic black holes—if ever produced or detected—could reveal deviations from standard evaporation models. If evaporation behaves exactly as predicted by conventional models without any quantum corrections, the conjecture may require revision.

Gravitational Wave Signatures

Black holes are now routinely observed through gravitational waves by observatories such as the LIGO Scientific Collaboration. Some theoretical models predict that exotic structures inside black holes could produce gravitational wave echoes after mergers. If such echoes are never observed despite increasingly precise measurements, it may suggest that black holes contain no internal structure beyond classical predictions. This would weaken theories involving internal cosmogenic processes.

Consistency with Future Quantum Gravity Theories

A complete theory of quantum gravity will ultimately determine whether spacetime topology can change under extreme conditions. If future theories rigorously demonstrate that topological transitions are forbidden during gravitational collapse, then the core mechanism of the conjecture would be ruled out. Conversely, if such transitions are permitted, the conjecture would gain theoretical support.

Epilogue

Hello, dear readers.

The journey in this book ends here. What began as a simple curiosity slowly grew into a long exploration of ideas, questions, experiments, and possibilities. Every chapter of this book was written not only to explain physics but also to share the thought process behind discovering and understanding it. Before we close the final page, I would like to briefly reflect on how each chapter came into existence during this journey.

Chapter 1: This chapter was quite interesting for me as well. When I first began writing it, I myself did not know where exactly to place *The Watcher.* I had the concept of dimensions very clear to me, but the very question I hooked you with initially, I didn't know the answer to either. I struggled to decide how to introduce the idea. But as the writing progressed, the answer revealed itself through the flow of the narrative.

Chapter 2: I thought of writing this chapter while studying entropy and the concept of time dilation. And honestly, this chapter became one of the most dynamic parts of the book. It was constantly growing. Almost every day, I would think, *"Oh, I cannot miss that concept!"* and add something new. Because of that, this chapter saw frequent additions and revisions throughout the writing process. Even now, I am not completely sure whether I managed to include everything important. If you feel that something essential is missing, I sincerely ask for your forgiveness.

Chapter 3: This chapter turned out to be somewhat like a mini space encyclopedia. Writing it required a great deal of research. While studying different topics, I kept discovering fascinating ideas and phenomena that I had not initially planned to include. Many of those discoveries found their way into the chapter. In a sense, the chapter evolved along with my own learning process.

Chapter 4: Personally, this is one of my favourite chapters. In fact, it is one of the main reasons I began writing this book in the first place. Whenever someone asks me to intrigue them with physics, I immediately talk about the double-slit experiment with electrons. Few experiments capture the strange beauty of quantum mechanics the way this one does. I also wanted to explore the story behind the Higgs Boson, and that naturally led to introducing elements of particle physics in the chapter.

Chapter 5: If Chapter 4 was the first major reason for writing this book, this chapter was the second. I remember reading a random article online with the phrase *"Something created out of nothing."* At first, I could hardly believe it. The idea sounded almost unbelievable. But as I explored the physics behind it, I realised that it connects deeply with how modern theories describe the birth of particles and even the universe itself. I hope the same sense of amazement reached you as well.

Chapter 6: I often hear statements such as *mass is energy* or discussions about relativity in everyday science conversations. These ideas form the very core foundations of modern physics, yet they are not always explained clearly. While studying these theories myself, I felt the need to organise and clarify them. That became the motivation behind this chapter—to understand them better and to explain them to you in a simple way.

Chapter 7: Interestingly, when I first planned this chapter, I expected it to be the shortest chapter in the book. However, as my research continued, I encountered more and more information related to the topic. The deeper I went, the more fascinating details I discovered. By the time everything was compiled and explained, it had grown into the second-largest chapter in the entire book.

Chapter 8: This chapter was never part of the original plan. One morning, I suddenly realised that there were several fascinating topics—stories, theories, and ideas—that I simply could not ignore just because they did not perfectly fit into earlier chapters. They were too interesting to be left out. So this chapter was born as a kind of bonus collection of those concepts that I personally enjoyed exploring.

Chapter 9: This final chapter is something I still find difficult to describe.

The conjecture presented there did not appear suddenly in my mind like a *"Eureka!"* moment. Instead, it emerged slowly. I wanted to explore an idea that would be creative yet not violate known physical laws excessively. I spent days—sometimes weeks—thinking about possible ways to approach the problem of singularities.

The very first version of the idea was honestly quite absurd, and perhaps it is better that you never know what it was. Over time, however, the idea evolved and refined itself. I used artificial intelligence tools to test different possibilities and explore the mathematical consistency behind the concept. Gradually, the conjecture reached the form you see in the final chapter.

And that brings us to the end of this journey. Obviously, there is more which I can tell about, like how the writing process started, what my initial idea was, etc., etc. But, I think I've bored you with the Epilogue quite enough.

So, let me conclude with the thought that guided the entire process:

"Physics does not progress through equations.
It progresses through curiosity."

I hope your journey through these pages was both enjoyable and 'intriguing', obviously. And that the ideas explored here leave you with the same sense of wonder that inspired me.

After all, the universe still holds countless mysteries.

And somewhere among them may lie the next question worth asking...

About the Author

Ashmit is a Class 10 student at St. Thomas School, Ranchi, with a deep fascination for the mysteries of the universe. His interest in physics began in the 5th grade, when he started exploring questions about reality, space, and the fundamental nature of the cosmos. Particularly drawn to quantum physics, Ashmit enjoys thinking deeply about scientific ideas and learning through lectures, discussions, and exploration.

Through his book *Physics of the Intrigued,* he aims to make physics accessible to everyone: from curious beginners to serious enthusiasts, by explaining profound concepts in a simple and engaging way with minimal mathematics. The book also introduces one of his own theoretical conjectures inspired by his curiosity about the nature of reality. Ashmit writes to inspire readers to appreciate the beauty and mystery of the universe. Alongside physics, he enjoys programming, singing, watching shows, and exploring new places. Through his writing, Ashmit hopes to inspire readers, especially young minds, to ask deeper questions about the universe.

"Physics doesn't progress through equations. It progresses through curiosity,"
- ***Ashmit.***

You can connect with me on:

- https://www.instagram.com/planckscaleashmit
- mailto:rajashmit408@gmail.com

www.ingramcontent.com/pod-product-compliance
Lightning Source LLC
LaVergne TN
LVHW100306210926
848480LV00003B/1031